SpringerBriefs in Mathematics

SpringerBriefs in Mathematics showcases expositions in all areas of mathematics and applied mathematics. Manuscripts presenting new results or a single new result in a classical field, new field, or an emerging topic, applications, or bridges between new results and already published works, are encouraged. The series is intended for mathematicians and applied mathematicians.

More information about this series at http://www.springer.com/series/10030

Ping Zhang

Color-Induced Graph Colorings

 Springer

Ping Zhang
Department of Mathematics
Western Michigan University
Kalamazoo, MI, USA

ISSN 2191-8198   ISSN 2191-8201   (electronic)
SpringerBriefs in Mathematics
ISBN 978-3-319-20393-5   ISBN 978-3-319-20394-2   (eBook)
DOI 10.1007/978-3-319-20394-2

Library of Congress Control Number: 2015942698

Mathematics Subject Classification (2010): 05C15, 05C70, 05C78

Springer Cham Heidelberg New York Dordrecht London

Printed on acid-free paper

Springer International Publishing AG Switzerland is part of Springer Science+Business Media (www.
springer.com)

Preface

The interest in edge colorings of graphs can be traced back to 1880 when the Scottish mathematician Peter Guthrie Tait attempted to solve the Four Color Problem with the aid of edge colorings. Despite the fact that Tait's approach was not successful, it initiated a new concept. In 1964, Vadim Vizing proved that the minimum number of colors needed to color the edges of a graph so that every two adjacent edges are colored differently (proper edge colorings) is one of two numbers, namely either the maximum degree or the maximum degree plus one. This result led to an increased interest and study of edge colorings in graph theory, not only edge colorings that are proper but also edge colorings that are not.

In recent decades, there has been great interest in edge colorings that give rise to vertex colorings in a variety of ways, which is the subject of this book. While we will be describing many ways in which edge colorings have induced vertex colorings and some of the major results, problems, and conjectures that have arisen in this area of study, it is not our goal to give a detailed survey of these subjects. Indeed, it is our intention to provide an organized summary of several recent coloring concepts and topics that belong to this area of study, with the hope that this may suggest new avenues of research topics.

In Chap. 1, we begin with a brief review of the well-known concepts of proper edge colorings and proper vertex colorings, including many fundamental results concerning them.

In Chap. 2, unrestricted edge colorings of graphs are considered whose colors are elements of the set $\mathbb{N}$ of positive integers or a set $[k] = \{1, 2, \ldots, k\}$ for some positive integer k. From such an edge coloring c of a graph G, a sum-defined vertex coloring c' is defined, that is, for each vertex v of G, the color $c'(v)$ of v is the sum of the colors of the edges incident with v. The edge coloring c is vertex-distinguishing or irregular if the resulting vertex coloring c' has the property that $c'(u) \neq c'(v)$ for every pair u, v of distinct vertices of G. The minimum positive integer k for which a graph G has such a vertex-distinguishing edge coloring is the irregularity strength of G. In Chap. 3, the corresponding coloring is considered in which the colors are taken from a set $\mathbb{Z}_k$ of integers modulo k.

Chapter 4 also deals with unrestricted edge colorings c of graphs but here the induced vertex coloring is defined so that the color $c'(v)$ of a vertex v is the set of colors of its incident edges. In Chap. 5, the emphasis changes from vertex colorings that are set-defined to those that are multiset-defined. In both cases, the induced vertex colorings c' are vertex-distinguishing.

In Chap. 6, unrestricted edge colorings c of graphs are once again considered but in this case the induced vertex colorings c' are neighbor-distinguishing, that is, $c'(u) \neq c'(v)$ for every two adjacent vertices u and v. In this chapter, two vertex colorings c are defined, both where the colors belong to a set $[k]$, one where $c'(v)$ is sum-defined and the other where $c'(v)$ is multiset-defined. Chapter 7 is devoted to unrestricted edge colorings of graphs whose colors are elements of $\mathbb{Z}_k$ of integers modulo k that induce a sum-defined, neighbor-distinguishing vertex coloring.

In Chap. 8, both proper and unrestricted edge colorings are considered, and the vertex colorings are set-defined, using elements of $[k]$ as colors. In Chap. 9, the edge colorings are proper and the vertex colorings considered are sum-defined, using elements of $[k]$ as colors. In these two chapters, the properties of being vertex-distinguishing and neighbor-distinguishing are both described. Chapter 9 ends with a discussion of so-called twin edge colorings, which are proper edge colorings that use the elements of $\mathbb{Z}_k$ as colors and that induce proper vertex colorings that are sum-defined.

The following table summarizes all types of edge colorings considered in this book and the resulting vertex colorings. In particular, the table describes, in each chapter:

1. the condition placed on the edge coloring,
2. the sets from which the edge colors are selected,
3. the definition of the vertex colors, and
4. the property required of the resulting vertex coloring.

Chapter 1: Introduction

Chapter 2: The Irregularity Strength of a Graph

Unrestricted Edge Colorings, $\mathbb{N}$, Sum-defined, Vertex-Distinguishing.

Chapter 3: Modular Sum-defined, Irregular Colorings

Unrestricted Edge Colorings, $\mathbb{Z}_k$, Sum-defined, Vertex-Distinguishing.

Chapter 4: Set-Defined Irregular Colorings

Unrestricted Edge Colorings, $\mathbb{N}$, Set-defined, Vertex-Distinguishing.

Chapter 5: Multiset-Defined Irregular Colorings

Unrestricted Edge Colorings, $\mathbb{N}$, Multiset-defined, Vertex-Distinguishing.

Chapter 6: Sum-Defined Neighbor-Distinguishing Colorings

Unrestricted Edge Colorings, $\mathbb{N}$, Sum-defined, Neighbor-Distinguishing.

Chapter 7: Modular Sum-Defined Neighbor-Distinguishing Colorings

Unrestricted Edge Colorings, $\mathbb{Z}_k$, Sum-defined, Neighbor-Distinguishing.

Kalamazoo, MI, USA   Ping Zhang

Acknowledgements

With pleasure, the author thanks Gary Chartrand for the advice and information he kindly supplied on many topics described in this book. In addition, the author thanks the reviewers for the valuable input and suggestions they provided with the first draft of this manuscript. Finally, the author is very grateful to Razia Amzad, SpringerBriefs editor, for her kindness and encouragement in writing this book. It is because of all of you that an improved book resulted.

Contents

List of Figures

Chapter 1
Introduction

One of the most popular areas of study in graph theory is colorings. This topic can be traced back to the origin of the Four Color Problem and whether it is possible to color the regions of every map with four or fewer colors in such a way that every two regions having a common boundary are assigned different colors. Later it was seen that this problem could be looked at as a problem in graph theory—whether it is always possible to color the regions of every planar graph (embedded in the plane) so that every two adjacent regions are colored differently. It became known that the Four Color Problem could be solved if it could be solved for all bridgeless cubic planar graphs.

1.1 The Origin of Edge Colorings

The Scottish mathematician Tait [76] discovered a unique approach to solve the Four Color Problem. He proved that the edges of a bridgeless cubic planar graph G can be colored with three colors so that every two adjacent edges are colored differently if and only if the regions of G can be colored with four colors so that every two adjacent regions are colored differently. Although Tait's approach never led to a solution of the Four Color Problem, he was able to prove that such a 3-coloring of the edges of G induce an appropriate 4-coloring of the regions of G. The goal of this book is to describe a variety of edge colorings that have been introduced which induce, in a number of ways, vertex colorings possessing desirable properties.

Colors can be objects of any type. While initially, the colors that were used to color the regions of maps were actual colors such as red, blue, green and so on, later it became common to use positive integers for colors as these were simpler and it was easier to keep track of the number of colors being used. Later yet, elements of $\mathbb{Z}_k$, the integers modulo k, for some $k \geq 2$, were sometimes used as colors. Subsets or multisets of some set were also used as colors. Originally, the only requirement

P. Zhang, *Color-Induced Graph Colorings*, SpringerBriefs in Mathematics,
DOI 10.1007/978-3-319-20394-2_1

for assigning colors to the edges of a graph was that adjacent edges were required to be colored differently, resulting in *proper edge colorings*. Later, other restrictions were placed on edge colorings of graphs such as having distinct colors for its edges (so-called *rainbow colorings*) or a single color (a *monochromatic coloring*).

In Tait's theorem that there exists a proper 4-coloring of the regions of a bridgeless cubic planar graph if and only if there exists a proper 3-coloring of the edges of the graph, a proof that a 4-coloring of the regions of such a graph can result in a 3-coloring of its edges can be constructed by coloring the regions with the elements of the Klein four-group $\mathbb{Z}_2 \times \mathbb{Z}_2$ and then assigning the color to an edge of G which is the sum of the colors of its two incident regions. Since these colors are distinct, the edges are colored with the three nonzero elements of $\mathbb{Z}_2 \times \mathbb{Z}_2$.

Tait's theorem can be considered as the beginning of a class of problems in which some type of coloring in a graph leads to another type of coloring in the graph. In this book we describe numerous edge colorings that have given rise to vertex colorings in some manner. For the most part, the edge colorings that we will encounter are unrestricted, that is, no condition is placed on the manner in which the edges may be colored. In Chaps. 8 and 9, however, it is the most popular type of edge colorings with which we will be dealing, namely proper edge colorings.

The vertex colorings that are generated from edge colorings will have one of two properties. The vertex colorings will either be proper or vertex-distinguishing (also called a rainbow coloring).

Because proper edge colorings and proper vertex colorings are the most studied graph colorings and because it is results dealing with these colorings to which we will most often be referring, this chapter reviews these two coloring concepts as well as some of the theorems that have been obtained about them. We refer to the books [19, 22] for graph theoretic notation and terminology not described in this work.

1.2 Proper Edge Colorings

A *proper edge coloring* c of a nonempty graph G (a graph with edges) is a function $c : E(G) \rightarrow S$, where S is a set of colors (and so $S = [k]$ or $S = \mathbb{Z}_k$), with the property that $c(e) \neq c(f)$ for every two adjacent edges e and f of G. If the colors are chosen from a set of k colors, then c is called a *k-edge coloring* of G. The minimum positive integer k for which G has a k-edge coloring is called the *chromatic index* of G and is denoted by $\chi'(G)$.

It is immediate for every nonempty graph G that $\chi'(G) \geq \Delta(G)$, where $\Delta(G)$ is the maximum degree of G. The most important theorem dealing with chromatic index is one obtained by the Russian mathematician Vadim Vizing.

Theorem 1.1 ([78]). *For every nonempty graph G,*

$$\chi'(G) \leq \Delta(G) + 1.$$

As a result of Vizing's theorem, the chromatic index of every nonempty graph G is one of two numbers, namely $\Delta(G)$ or $\Delta(G) + 1$. A graph G with $\chi'(G) = \Delta(G)$ is called a *class one graph* while a graph G with $\chi'(G) = \Delta(G) + 1$ is called a *class two graph*. The chromatic index of complete graphs is given in the following result.

Theorem 1.2. *For each integer $n \geq 2$,*

$$\chi'(K_n) = \begin{cases} n - 1 & \text{if } n \text{ is even} \\ n & \text{if } n \text{ is odd.} \end{cases}$$

Therefore, K_n is a class one graph if n is even and is a class two graph if n is odd. The fact that K_n is a class one graph if and only if n is even is also a consequence of the following.

Theorem 1.3. *A regular graph G is a class one graph if and only if G is 1-factorable.*

An immediate consequence of this result is stated next.

Corollary 1.4. *Every regular graph of odd order is a class two graph.*

The next two results describe classes of graphs that are class one graphs. The first theorem is due to Denés König.

Theorem 1.5 ([69]). *Every bipartite graph is a class one graph.*

The following result is due to Jean-Claude Fournier.

Theorem 1.6 ([43]). *Let G be an nonempty graph. If the subgraph of G induced by the vertices of degree $\Delta(G)$ is a forest, then G is a class one graph.*

From this, we have the following corollary.

Corollary 1.7 ([43]). *If G is a graph in which no two vertices of maximum degree are adjacent, then G is a class one graph.*

If a graph G of odd order has sufficiently many edges, then G must be a class two graph. A graph G of order n and size m is called *overfull* if $m > \Delta(G)\lfloor n/2 \rfloor$. If G has even order n, then $m \leq \Delta(G)\lfloor n/2 \rfloor$ and so G is not overfull. On the other hand, a graph of odd order may be overfull.

Theorem 1.8. *Every overfull graph is a class two graph.*

1.3   Proper Vertex Colorings

A *proper vertex coloring* of a graph G is a function $c' : V(G) \rightarrow S$, where in our case, $S \subseteq \mathbb{N}$ or $S = \mathbb{Z}_k$ for some integer $k \geq 2$ such that $c'(u) \neq c'(v)$ for every pair u, v of adjacent vertices of G. If $|S| = k$, then c' is called a *k-vertex coloring* (or, more often, simply a *k-coloring*) of G. The minimum positive integer k for which G has a k-vertex coloring is called the *chromatic number* of G, denoted by $\chi(G)$.

For graphs of order $n \geq 3$, it is immediate which graphs of order n have chromatic number $1, n$ or 2.

Observation 1.9. *If G is a graph of order $n \geq 3$, then*

(a) $\chi(G) = 1$ *if and only if G is empty.*
(b) $\chi(G) = n$ *if and only if $G = K_n$.*
(c) $\chi(G) = 2$ *if and only if G is a nonempty bipartite graph.*

An immediate consequence of Observation 1.9(c) is that $\chi(G) \geq 3$ if and only if G contains an odd cycle. One of the most useful lower bounds for the chromatic number of a graph is stated below.

Proposition 1.10. *If H is a subgraph of a graph G, then $\chi(H) \leq \chi(G)$.*

The *clique number* $\omega(G)$ of a graph G is the maximum order of a complete subgraph of G. The following result is therefore a consequence of Proposition 1.10.

Corollary 1.11. *For every graph G, $\omega(G) \leq \chi(G)$.*

By Corollary 1.11 (or, in fact, by Observation 1.9(c)), if a graph G contains a triangle, then $\chi(G) \geq 3$. As the following result (proved by many) indicates, there are graphs G for which the lower bound for $\chi(G)$ and the clique number $\omega(G)$ of G may differ significantly.

Theorem 1.12. *For every integer $k \geq 2$, there exists a triangle-free graph G with $\chi(G) = k$.*

As far as upper bounds for the chromatic number of a graph are concerned, the following result gives such a bound in terms of the maximum degree of the graph.

Theorem 1.13. *For every graph G, $\chi(G) \leq \Delta(G) + 1$.*

For each odd integer $n \geq 3$, the connected graphs C_n and K_n have the property that $\chi(C_n) = 3 = \Delta(C_n) + 1$ and $\chi(K_n) = n = \Delta(K_n) + 1$. The British mathematician Rowland Leonard Brooks showed that these two classes of graphs are the only connected graphs with this property.

Theorem 1.14 ([15]). *If G is a connected graph that is neither an odd cycle nor a complete graph, then $\chi(G) \leq \Delta(G)$.*

Chapter 2
The Irregularity Strength of a Graph

Throughout Chaps. 2–7, we will be concerned with connected graphs G of order $n \geq 3$ and size m and an unrestricted edge coloring of G, that is, no condition is placed on the manner in which colors are assigned to the edges of G.

The unrestricted edge colorings inducing vertex colorings that have attracted the most attention are those where the vertex colorings are either vertex-distinguishing or neighbor-distinguishing. In this chapter, we consider a particular example of the first of these.

A nontrivial graph has been called *irregular* if its vertices have distinct degrees. It is well known that there is no such graph; that is, no graph is irregular. This observation led to a concept introduced by Gary Chartrand at the 250th Anniversary of Graph Theory Conference held at Indiana University-Purdue University Fort Wayne in 1986.

For a connected graph G, a *weighting* w of G is an assignment of numbers (usually positive integers) to the edges of G, where $w(e)$ denotes the weight of an edge e of G. This then converts G into a weighted graph in which the (*weighted*) *degree* of a vertex v is defined as the sum of the weights of the edges incident with v. A weighted graph G is then *irregular* if the vertices of G have distinct degrees. Later this concept was viewed in another setting.

2.1 Sum-Defined Vertex Colorings: Irregularity Strength

Rather than consider connected graphs G of order at least 3 whose edges are assigned weights, resulting in irregular weighted graphs, we can view this as *vertex-distinguishing* edge colorings of G where the induced vertex coloring is sum-defined and where then the vertices of G have distinct colors. Such vertex colorings are also referred to as *rainbow vertex colorings*.

© Ping Zhang 2015
P. Zhang, *Color-Induced Graph Colorings*, SpringerBriefs in Mathematics,
DOI 10.1007/978-3-319-20394-2_2

We now formally define such vertex-distinguishing edge colorings. Let $\mathbb{N}$ denote the set of positive integers and let E_v denote the set of edges incident with a vertex v in a graph G. An unrestricted edge coloring $c : E(G) \to \mathbb{N}$ induces a vertex coloring $c' : V(G) \to \mathbb{N}$, defined by

$$c'(v) = \sum_{e \in E_v} c(e) \quad \text{for each vertex } v \text{ of } G. \tag{2.1}$$

Proposition 2.1. *Let G be a nontrivial connected graph and let $c : E(G) \to \mathbb{N}$ be an edge coloring of G, where $c' : V(G) \to \mathbb{N}$ is the induced vertex coloring defined in (2.1). Then there exists an even number of vertices of odd color.*

Proof. Let $E(G) = \{e_1, e_2, \ldots, e_m\}$. Since

$$\sum_{v \in V(G)} c'(v) = 2 \sum_{i=1}^{m} c(e_i)$$

is even, there exists an even number of vertices of odd color. $\square$

While no edge coloring of the graph K_2 can induce a rainbow vertex coloring defined in this manner, there is a vertex-distinguishing edge coloring for every connected graph G of order 3 or more. To see this, let $E(G) = \{e_1, e_2, \ldots, e_m\}$ where then $m \geq 2$ and let c be the edge coloring of G defined by $c(e_i) = 2^{i-1}$ for $1 \leq i \leq m$. Since no two vertices are incident with the same set of edges, c induces a rainbow vertex coloring. This edge coloring shows that there is always a vertex-distinguishing edge coloring of a connected graph of size $m \geq 2$ where the largest color used is 2^{m-1}. In general, there exist vertex-distinguishing edge colorings of a graph of size m whose largest color is considerably less than 2^{m-1}.

For a connected graph G of size $m \geq 2$, the minimum of the largest colors used among the vertex-distinguishing edge colorings of G is called the *irregularity strength* of G and is denoted by $s(G)$. (The *strength* of a multigraph M is the maximum number of parallel edges joining two vertices of M.) Therefore, for a connected graph G of order at least 3, there exists an edge coloring $c : E(G) \to [k] = \{1, 2, \ldots, k\}$ for every integer k with $k \geq s(G)$ such that the induced (sum-defined) vertex coloring c' is vertex-distinguishing but there is no such edge coloring $c : E(G) \to [k]$ with this property for any integer k with $1 \leq k < s(G)$.

Since no nontrivial graph is irregular, it follows that every connected graph of order at least 3 must have irregularity strength at least 2. It is well known that there is exactly one connected graph G_n of order n for each $n \geq 2$ containing exactly two vertices having the same degree. All of these graphs have irregularity strength 2.

Proposition 2.2. *If G_n is the unique connected graph of order $n \geq 3$ containing exactly two vertices of equal degree, then $s(G_n) = 2$.*

Proof. As mentioned above, $s(G_n) \geq 2$ for every integer $n \geq 2$. Each such graph G_n can be described as having vertex set $V(G_n) = \{v_1, v_2, \ldots, v_n\}$ where $v_i v_j \in E(G_n)$ if and only if $i + j \leq n + 1$. Consequently,

$$\deg v_i = \begin{cases} n - i & \text{if } 1 \leq i \leq \lceil n/2 \rceil \\ n + 1 - i & \text{if } \lceil n/2 \rceil + 1 \leq i \leq n. \end{cases} \qquad (2.2)$$

So $\deg v_{\lceil \frac{n}{2} \rceil} = \deg v_{\lceil \frac{n}{2} \rceil + 1} = \lfloor n/2 \rfloor$. Let c be the edge coloring of G_n in which each edge is assigned the color 2 except for $v_1 v_{\lceil \frac{n}{2} \rceil + 1}$, which is colored 1. Then

$$c'(v_i) = \begin{cases} 2 \deg v_i & \text{if } i \neq 1, \lceil n/2 \rceil + 1 \\ 2 \deg v_i - 1 & \text{if } i = 1, \lceil n/2 \rceil + 1. \end{cases}$$

Since c' is vertex-distinguishing, $s(G) \leq 2$ and so $s(G) = 2$. $\square$

To show that every complete graph of order $n \geq 3$ has irregularity strength 3, we first make an observation concerning the irregularity strength of every regular graph.

Proposition 2.3. *The irregularity strength of every regular graph of order 3 or more is at least 3.*

Proof. Suppose that there exists an edge coloring of a regular graph G of order at least 3 with the colors 1 and 2 and that H is the spanning subgraph of G whose edges are color 1. Then H has two vertices u and v of equal degree. Since u and v have the same induced color in G, it follows that $s(G) \geq 3$. $\square$

Theorem 2.4 ([24]). *For each integer $n \geq 3$, $s(K_n) = 3$.*

Proof. By Proposition 2.3, it follows that $s(K_n) \geq 3$. To establish the inequality $s(K_n) \leq 3$, we show that there is a vertex-distinguishing edge coloring of K_n with the colors 1, 2 and 3. Since the edge coloring of K_3 given in Fig. 2.1 has this property, we may assume that $n \geq 4$.

Let G_n be the unique connected graph of order $n \geq 4$ having exactly two vertices of equal degree that is described in the proof of Theorem 2.2. Thus $V(G_n) = \{v_1, v_2, \ldots, v_n\}$ whose degrees are given in (2.2). As noted there, these equal degrees are $\lfloor n/2 \rfloor$. Assign the color 2 to the edges of G_n and the color 1 to the edges of its complement $\overline{G}_n$. The induced vertex colors $c^*(v_i)$ for this edge coloring of K_n are then

$$c^*(v_i) = 2 \deg_{G_n} v_i + (n - 1 - \deg_{G_n} v_i) = n - 1 + \deg_{G_n} v_i \qquad (2.3)$$

Fig. 2.1 Showing $s(K_3) = 3$

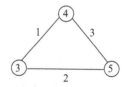

for $1 \le i \le n$. Next, increase the color of each of the edges $v_1 v_2, v_1 v_3, \ldots, v_1 v_{\lceil \frac{n}{2} \rceil}$ by 1, resulting in an edge coloring c using the colors 1, 2, 3. By (2.3), the induced vertex coloring c' of K_n satisfies

$$c'(v_i) = \begin{cases} (2n - 2) + (\lceil n/2 \rceil - 1) & \text{if } i = 1 \\ n + \deg_{G_n} v_i & \text{if } 2 \le i \le \lceil n/2 \rceil \\ (n - 1) + \deg_{G_n} v_i & \text{if } \lceil n/2 \rceil + 1 \le i \le n. \end{cases}$$

It then follows by (2.2) that

$$c'(v_i) = \begin{cases} 2n + \lceil n/2 \rceil - 3 & \text{if } i = 1 \\ 2n - i & \text{if } 2 \le i \le n. \end{cases}$$

Since the revised edge coloring c of K_n is vertex-distinguishing, $s(K_n) \le 3$ and so $s(K_n) = 3$. $\square$

2.2   On the Irregularity Strength of Regular Graphs

We saw in Proposition 2.3 that the irregularity strength of every regular graph of order 3 or more is at least 3 and in Theorem 2.4 that the irregularity strength of the complete graph K_n, $n \ge 3$, an $(n - 1)$-regular graph, is 3. We now investigate the irregularity strength of regular graphs in more detail. First, we present a lower bound for the irregularity strength of a graph G in terms of the number of vertices of a specific degree in G.

Proposition 2.5 ([24]). *Let G be a connected graph of order $n \ge 3$ with minimum degree $\delta(G)$ and maximum degree $\Delta(G)$ containing n_i vertices of degree i for each integer i with $\delta(G) \le i \le \Delta(G)$. Then*

$$s(G) \ge \max \left\{ \frac{n_i - 1}{i} + 1 : \delta(G) \le i \le \Delta(G) \right\}.$$

Proof. Suppose that $s(G) = s$. Let there be given a vertex-distinguishing edge coloring of G with the colors $1, 2, \ldots, s$ and let $v \in V(G)$ where $\deg v = i$. Then the induced vertex color $c'(v)$ satisfies $i \le c'(v) \le si$. Hence each vertex of degree i has one of the $si - i + 1 = i(s - 1) + 1$ induced colors in the set $\{i, i + 1, \ldots, si\}$ and so $n_i \le i(s - 1) + 1$. Therefore,

$$s(G) = s \ge \frac{n_i - 1}{i} + 1$$

for each i with $\delta(G) \le i \le \Delta(G)$. $\square$

If G is a regular graph, then Proposition 2.5 has the following corollary.

Corollary 2.6 ([24]). *If G is a connected r-regular graph, $r \geq 2$, of order $n \geq 3$, then*

$$s(G) \geq \frac{n-1}{r} + 1.$$

When $n \equiv 2 \pmod 4$ or $n \equiv 3 \pmod 4$, Corollary 2.6 can be improved a bit.

Corollary 2.7 ([24]). *If G is a connected r-regular graph of order $n \geq 3$ where $n \equiv 2 \pmod 4$ or $n \equiv 3 \pmod 4$, then*

$$s(G) > \frac{n-1}{r} + 1.$$

Proof. Suppose that $n \equiv 2 \pmod 4$ and assume, to the contrary, that $s(G) = s = \frac{n-1}{r} + 1$. Then there is a vertex-distinguishing edge coloring of G with the colors $1, 2, \ldots, s$. Hence each induced vertex color is one of the $sr - r + 1$ colors $r, r + 1, \ldots, sr$. By assumption, $n = sr - r + 1$ and so the induced vertex colors are precisely the n colors $r, r + 1, \ldots, sr$. However, $n/2$ of these colors are odd, that is, G has an odd number of vertices of odd color, contradicting Proposition 2.1.

The argument when $n \equiv 3 \pmod 4$ is similar.   □

By Corollary 2.7, the irregularity strength of the Petersen graph P satisfies $s(P) > \frac{10-1}{3} + 1 = 4$, that is, $s(P) \geq 5$. Since the edge coloring of the Petersen graph with the colors $1, 2, \ldots, 5$ shown in Fig. 2.2 is vertex-distinguishing, $s(P) \leq 5$ and so $s(P) = 5$.

Since, by Theorem 2.4, $s(K_n) = 3$ for every integer $n \geq 3$, it follows that the complete n-partite graph in which every partite set consists of a single vertex has irregularity strength 3. We now see that this is also true when each partite set consists of exactly two vertices. For each integer $r \geq 2$, we write $K_{r(2)}$ for the $(2r-2)$-regular complete r-partite graph where each partite set consists of two vertices.

Theorem 2.8 ([52]). *For each integer $r \geq 2$, $s(K_{r(2)}) = 3$.*

Fig. 2.2 An edge coloring of the Petersen graph

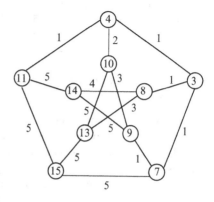

Proof. Since it is easy to see that $s(C_4) = 3$ and $K_{2(2)} = C_4$, we may assume that $r \geq 3$. Let $G = K_{r(2)}$. Since G is a $(2r - 2)$-regular graph of order $2r$, it follows by Corollary 2.6 that $s(G) \geq 3$. We show that $s(G) \leq 3$ by describing a vertex-distinguishing edge coloring $c : E(G) \to \{1, 2, 3\}$.

Denote the partite sets of G by $V_1, V_2, \ldots, V_r$, where $V_i = \{x_i, y_i\}$ for $1 \leq i \leq r$. We now relabel the vertices of G by $u_1, u_2, \ldots, u_n$, where $n = 2r$, such that $u_i = x_i$ for $1 \leq i \leq r$ and $u_{n+1-i} = y_i$ for $1 \leq i \leq r$. Let H be the spanning subgraph of G where $u_i u_j \in E(H)$ if $1 \leq i < j \leq n$ and $i + j \leq n$. Thus

$$\deg_H u_i = \begin{cases} 2r - 1 - i & \text{if } 1 \leq i \leq r \\ 2r - i & \text{if } r + 1 \leq i \leq n. \end{cases} \tag{2.4}$$

Thus $\deg_H u_1 \geq \deg_H u_2 \geq \cdots \geq \deg_H u_n$ and $\deg_H u_i = \deg_H u_{i+1}$ only when $i = r$. Next, we define an edge coloring $\bar{c} : E(G) \to \{1, 2, 3\}$ of G by assigning the color 1 to each edge of H and the color 3 to the remaining edges of G. The induced vertex coloring $\bar{c}'$ is then defined by

$$\bar{c}'(u_i) = \deg_H u_i + 3(2r - 2 - \deg_H u_i) = 6r - 6 - 2\deg_H u_i$$

for $1 \leq i \leq n$. Hence $\bar{c}'(u_1) \leq \bar{c}'(u_2) \leq \cdots \leq \bar{c}'(u_n)$ with equality only for $\bar{c}'(u_r)$ and $\bar{c}'(u_{r+1})$. In particular, $\bar{c}'(u_r) = \bar{c}'(u_{r+1}) = 4r - 4$.

We now revise the edge coloring $\bar{c}$ by replacing the color 1 of $u_1 u_r$ by 2, producing a new edge coloring c of G. The induced vertex coloring c' then satisfies the following

$$c'(u_i) = \begin{cases} 2r - 1 & \text{if } i = 1 \\ 6r - 6 - 2\deg_H u_i & \text{if } 2 \leq i \leq r - 1 \text{ or } r + 1 \leq i \leq n \\ 4r - 3 & \text{if } i = r. \end{cases}$$

This is illustrated for $K_{4(2)}$ in the following table.

$u_1, u_2, \ldots, u_8$	x_1	x_2	x_3	x_4	y_4	y_3	y_2	y_1
$\deg_H u_i$	6	5	4	3	3	2	1	0
$\bar{c}'(u_i)$	6	8	10	12	12	14	16	18
$c'(u_i)$	7	8	10	13	12	14	16	18

Since c is a vertex-distinguishing edge coloring, it follows that $s(G) \leq 3$ and so $s(G) = 3$. $\qquad \square$

Even though each complete multipartite graph in which every partite set consists of exactly one vertex or every partite set consists of exactly two vertices has irregularity strength 3, this is not the case if every partite set consists of exactly three vertices, as we now illustrate with the graph $K_{3,3}$. By Corollary 2.7, $s(K_{3,3}) \geq 3$. Assume to the contrary that $s(K_{3,3}) = 3$. Then there is a vertex-distinguishing edge coloring c of $G = K_{3,3}$ with induced vertex coloring c'. Therefore, $\{c'(v) : v \in V(G)\} \subseteq S = \{3, 4, \ldots, 9\}$. Since the order of G is 6 and $|S| = 7$, every integer in

S is a vertex color of G except for one color in S. Because S consists of four odd integers and three even integers and every graph has an even number of vertices of odd color (by Proposition 2.1), each of the integers $3, 5, 7, 9$ is the color of exactly one vertex of G. Suppose that $c'(x) = 3$ and $c'(y) = 9$. Then the three edges incident with x are colored 1 and the three edges incident with y are colored 3. This implies that x and y belong to the same partite set U of G. Thus each vertex belonging to the other partite set W of G is incident with at least one edge colored 1 and at least one edge colored 3. Thus, the colors of the three vertices in W are 5, 6 and 7. Since the sum of the colors of the three vertices of W is 18, the the sum of the colors of the three vertices of U is also 18, which implies that the colors of the three vertices in U are 3, 6 and 9. This is impossible, however, since there is a vertex of W colored 6. Therefore, $s(K_{3,3}) \geq 4$. We now show that not only $s(K_{3,3}) = 4$ but provide information about the value of $s(K_{r,r})$ for every integer $r \geq 2$.

For two disjoint subsets A and B of the vertex set of a graph G, let $[A, B]$ denote the set of edges joining a vertex of A and a vertex of B.

Theorem 2.9 ([24, 51]). *For an integer $r \geq 2$,*

$$s(K_{r,r}) = \begin{cases} 3 & \text{if } r \text{ is even} \\ 4 & \text{if } r \text{ is odd.} \end{cases}$$

Proof. Denote the partite sets of $G = K_{r,r}$ by

$$U = \{u_1, u_2, \ldots, u_r\} \text{ and } W = \{w_1, w_2, \ldots, w_r\}.$$

By Corollary 2.6, $s(G) \geq 3$. Assume first that r is even. Then $r = 2k$ for some integer k. Define an edge coloring $c : E(G) \to \{1, 2, 3\}$ by

$$c(u_i w_j) = \begin{cases} 1 & \text{if } j > i \text{ or } i = j \geq k + 1 \\ 2 & \text{if } i = j \leq k \\ 3 & \text{if } j < i. \end{cases}$$

Then the induced vertex coloring c' satisfies the following

$$c'(u_i) = \begin{cases} r + (2i - 1) & \text{if } 1 \leq i \leq k \\ r - 2 + 2i & \text{if } k + 1 \leq i \leq 2k \end{cases}$$

$$c'(w_i) = \begin{cases} 3r + 1 - 2i & \text{if } 1 \leq i \leq k \\ 3r - 2i & \text{if } k + 1 \leq i \leq 2k. \end{cases}$$

Consequently, $c' : V(G) \to \{r, r + 1, \ldots, 3r - 1\}$ is vertex-distinguishing. The colorings c and c' are illustrated for $K_{4,4}$ in Fig. 2.3. Since c is a vertex-distinguishing edge coloring, it follows that $s(G) \leq 3$ and so $s(G) = 3$ if r is even.

Fig. 2.3 An edge coloring of $K_{4,4}$

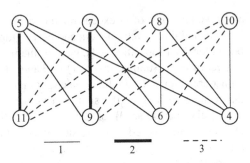

Next, assume that $r \geq 3$ is odd. First, we show that $s(G) \geq 4$. Assume, to the contrary, that $s(G) = 3$. Then there exists an edge coloring $\bar{c} : E(G) \rightarrow \{1, 2, 3\}$ such that $\bar{c}' : V(G) \rightarrow \{r, r+1, \ldots, 3r\} = T$ is vertex-distinguishing. Since $|T| = 2r + 1$ and there is an even number of vertices of odd color, there is an even integer $t \in T$ that is not the color of any vertex in G.

If 1 is subtracted from each edge color, then we obtain a vertex-distinguishing edge coloring $c : E(G) \rightarrow \{1, 2\}$ such that $c' : V(G) \rightarrow \{0, 1, \ldots, 2r\}$. Hence the odd color $i = t - r$ is not the color of any vertex of G.

Let $V(G) = S \cup L$, where $|S| = |L| = r$, such that S is the set of vertices of G having the smallest r colors and L is the set of vertices of G having the largest r colors. Let

$$\sigma(S, L) = \sum_{e \in [S, L]} c(e),$$

$U_L = U \cap L$, $W_L = W \cap L$, $a = |U_L|$ and $b = |W_L|$. Then $a + b = r$. If $x \in U_L$, then $\sum_{e \in [\{x\}, W_L]} c(e) \leq 2b$; while if $x \in W_L$, then $\sum_{e \in [\{x\}, U_L]} c(e) \leq 2a$. Therefore,

$$\sigma(S, L) \geq \sum_{x \in U_L} [c'(x) - 2b] + \sum_{x \in W_L} [c'(x) - 2a] = \left[\sum_{x \in L} c'(x) \right] - 4ab.$$

Since $a + b = r$ and r is odd, the maximum value of ab is $\frac{1}{4}(r^2 - 1)$. Hence

$$\sigma(S, L) \geq \left[\sum_{x \in L} c'(x) \right] - (r^2 - 1). \tag{2.5}$$

We consider two cases, according to whether $i \leq r$ or $i \geq r + 2$.

Case 1. $i \leq r$. Since $\{c'(x) : x \in L\} = \{r+1, r+2, \ldots, 2r\}$, it follows by (2.5) that

$$\sigma(S, L) \geq (r + 1 + r + 2 + \cdots + 2r) - (r^2 - 1) = \frac{r^2 + r + 2}{2}. \tag{2.6}$$

On the other hand, $\{c'(x) : x \in S\} = \{0, 1, 2, \ldots, r\} - \{i\}$ and the sum of these colors is maximum when $i = 1$. Thus,

$$\sigma(S, L) \leq 0 + 2 + 3 + \cdots + r = \frac{r^2 + r - 2}{2},$$

which contradicts (2.6).

Case 2. $i \geq r + 2$. Then $\{c'(x) : x \in L\} = \{r, r + 1, \ldots, 2r\} - \{i\}$ and the sum of these colors is minimum when $i = 2r - 1$. It then follows by (2.5) that

$$\sigma(S, L) \geq [r + (r + 1) + \cdots + (2r - 2) + 2r] - (r^2 - 1) = \frac{r^2 - r + 4}{2}. \qquad (2.7)$$

On the other hand, $\{c'(x) : x \in S\} = \{0, 1, 2, \ldots, r - 1\}$. Hence

$$\sigma(S, L) \leq 0 + 1 + 2 + \cdots + (r - 1) = \frac{r^2 - r}{2},$$

which contradicts (2.7). Therefore, $s(G) \geq 4$.

It remains to show that there is a vertex-distinguishing edge coloring $c : E(G) \rightarrow \{1, 2, 3, 4\}$. Since $r \geq 3$ is odd, $r = 2k + 1$ for some positive integer k. Define an edge coloring $c : E(G) \rightarrow \{1, 2, 3, 4\}$ by

$$c(u_iw_j) = \begin{cases} 1 & \text{if } j > i \text{ and } (i, j) \neq (k + 1, k + 2) \\ & \text{or } i = j = k + 1 \\ 2 & \text{if } i = j \leq k \text{ or } (i, j) = (k + 1, 2k + 1) \\ 3 & \text{if } i = j \geq k + 2 \\ 4 & \text{if } j < i. \end{cases}$$

Then the induced vertex coloring c' satisfies the following

$$c'(u_i) = \begin{cases} r - 2 + 3i & \text{if } 1 \leq i \leq k + 1 \\ r - 1 + 3i & \text{if } k + 2 \leq i \leq 2k + 1 \end{cases}$$

$$c'(w_i) = \begin{cases} 4r + 1 - 3i & \text{if } 1 \leq i \leq k \\ 4r - 3i & \text{if } i = k + 1 \\ 4r + 2 - 3i & \text{if } k + 2 \leq i \leq 2k \\ r + 3 & \text{if } i = 2k + 1. \end{cases}$$

The vertex coloring $c' : V(G) \rightarrow \{r, r + 1, \ldots, 4r\}$ is vertex-distinguishing. This is illustrated in Fig. 2.4 for $K_{5,5}$. Since c is a vertex-distinguishing edge coloring, it follows that $s(G) \leq 4$ and so $s(G) = 4$ if r is odd. $\square$

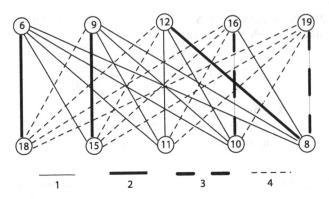

Fig. 2.4 An edge coloring of $K_{5,5}$

In [41], it was shown that if G is a regular complete k-partite graph where $k \geq 3$, then $s(G) = 3$. We now verify this statement by giving a proof along the same lines as the proofs of Proposition 2.2 and Theorems 2.4 and 2.8.

Theorem 2.10. *If G is a regular complete k-partite graph where $k \geq 3$, then*

$$s(G) = 3.$$

Proof. Let $G = K_{k(r)}$ where $k \geq 3$. Thus G is a $(k-1)r$-regular graph of order kr. By Proposition 2.3, $s(G) \geq 3$. Thus, it remains to show that G has a vertex-distinguishing edge coloring using the colors $1, 2, 3$. Let $V_1, V_2, \ldots, V_k$ denote the k partite sets of G where

$$V_i = \left\{ v_1^{(i)}, v_2^{(i)}, \ldots, v_r^{(i)} \right\} \quad \text{for } 1 \leq i \leq k.$$

First, suppose that r is even, say $r = 2\ell$ for some positive integer ℓ. We now construct an ordered list L of the n vertices of G, separated into r blocks $B_1, B_2, \ldots, B_r$ of k vertices each. The first block is $B_1 : v_1^{(1)}, v_1^{(2)}, \ldots, v_1^{(k)}$. In general, for $1 \leq j \leq \ell$, the block B_j is

$$B_j : v_j^{(1)}, v_j^{(2)}, \ldots, v_j^{(k)}. \tag{2.8}$$

For $\ell + 1 \leq j \leq r$, the block B_j is

$$B_j : v_j^{(k)}, v_j^{(k-1)}, \ldots, v_j^{(2)}, v_j^{(1)}. \tag{2.9}$$

Consequently, the list L is

$$L : B_1, B_2, \ldots, B_\ell, B_{\ell+1}, B_{\ell+2}, \ldots, B_r. \tag{2.10}$$

We relabel the vertices of L as $u_1, u_2, \ldots, u_n$. Next, we construct a spanning subgraph H of G as follows. For integers i and j with $1 \leq i < j \leq n$, the vertex u_i is adjacent to u_j in H if $i + j \leq n + 1$ and u_i and u_j do not belong to the same partite set of G. Thus $\deg_H u_1 \geq \deg_H u_2 \geq \cdots \geq \deg_H v_n$ and $\deg_H u_i = \deg_H u_{i+1}$

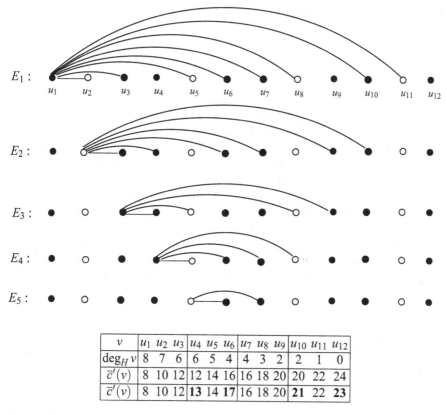

v	u_1	u_2	u_3	u_4	u_5	u_6	u_7	u_8	u_9	u_{10}	u_{11}	u_{12}
$\deg_H v$	8	7	6	6	5	4	4	3	2	2	1	0
$\overline{c}'(v)$	8	10	12	12	14	16	16	18	20	20	22	24
$\widetilde{c}'(v)$	8	10	12	13	14	17	16	18	20	21	22	23

Fig. 2.5 Constructing the graph H in $K_{3(4)}$

only when $i < n$ and $i \equiv 0 \pmod{k}$. For $G = K_{3(4)}$, the edge set $\cup_{i=1}^{5} E_5$ of the graph H is shown in Fig. 2.5 where $E_i = \{u_i u_j \in E(G) : i + j \le 13\}$ for $1 \le i \le 5$.

First, we define an edge coloring $\overline{c} : E(G) \to \{1, 3\}$ of G by assigning the color 1 to each edge of H and the color 3 to each edge in $G - E(H)$. The induced vertex coloring $\overline{c}' : V(G) \to \mathbb{N}$ satisfies the following:

(1) $\overline{c}'(u_i)$ is even for all i $(1 \le i \le n)$,
(2) $\overline{c}'(u_1) \le \overline{c}'(u_2) \le \cdots \le \overline{c}'(u_n)$ and
(3) $\overline{c}'(u_i) = \overline{c}'(u_{i+1})$ only when $i < n$ and $i \equiv 0 \pmod{k}$.

We now revise the edge coloring $\overline{c} : E(G) \to \{1, 3\}$ by constructing a new edge coloring $c : E(G) \to \{1, 2, 3\}$ as follows:

$$c(e) = \begin{cases} \overline{c}(e) + 1 & \text{if } e = u_{(j-1)k+1} u_{jk}, j \text{ even}, 2 \le j \le \ell, \\ \overline{c}(e) - 1 & \text{if } e = u_{(j-1)k+1} u_{jk}, j \text{ even}, \ell + 1 \le j \le 2\ell \\ \overline{c}(e) & \text{otherwise}. \end{cases}$$

Then the induced vertex coloring $c' : V(G) \to \mathbb{N}$ satisfies

$$c'(v) = \begin{cases} \overline{c}'(v) + 1 & \text{if } v = u_{(j-1)k+1}, u_{jk}, j \text{ even}, 2 \leq j \leq \ell \\ \overline{c}'(v) - 1 & \text{if } v = u_{(j-1)k+1}, u_{jk}, j \text{ even}, \ell + 1 \leq j \leq 2\ell \\ \overline{c}'(v) & \text{otherwise.} \end{cases}$$

It then follows by properties (1)–(3) of the vertex coloring $\overline{c}'$ that c' is vertex-distinguishing. This is also illustrated for $K_{3(4)}$ in Fig. 2.5.

Next, suppose that $r \geq 3$ is odd, say $r = 2\ell + 1$ for some positive integer ℓ. We now construct an ordered list L of the n vertices of G, separated into r blocks $B_1, B_2, \ldots, B_r$ of k vertices each. For $1 \leq j \leq \ell + 1$, the block B_j is the one in (2.8). For $\ell + 2 \leq j \leq r$, the block B_j is the one in (2.9). Consequently, the list L is as described in (2.10). Then relabel the vertices of L as $u_1, u_2, \ldots, u_n$. We now construct a spanning subgraph H of G as in the case when r is even. That is, for integers i and j with $1 \leq i < j \leq n$, the vertex u_i is adjacent to u_j in H if $i + j \leq n + 1$ and u_i and u_j do not belong to the same partite set of G. Thus $\deg_H u_1 \geq \deg_H u_2 \geq \cdots \geq \deg_H v_n$ and $\deg_H u_i = \deg_H u_{i+1}$ only when either

$$(1)\ i \equiv 0 \pmod{k} \text{ and } i \neq n, (\ell + 1)k \text{ or } (2)\ i = \left\lceil \frac{n}{2} \right\rceil.$$

First, we define an edge coloring $\overline{c} : E(G) \to \{1, 3\}$ of G by assigning the color 1 to each edge of H and the color 3 to each edge in $G - E(H)$. The induced vertex coloring $\overline{c}' : V(G) \to \mathbb{N}$ satisfies the following:

(1) $\overline{c}'(u_i)$ is odd for all i ($1 \leq i \leq n$) if $k - 1$ is odd and $\overline{c}'(u_i)$ is even ($1 \leq i \leq n$) if $k - 1$ is even,
(2) $\overline{c}'(u_1) \leq \overline{c}'(u_2) \leq \cdots \leq \overline{c}'(u_n)$ and
(3) $\overline{c}'(u_i) = \overline{c}'(u_{i+1})$ only when either $i \equiv 0 \pmod{k}$ and $i \neq n, (\ell + 1)k$ or $i = \left\lceil \frac{n}{2} \right\rceil$.

We now revise the edge coloring $\overline{c} : E(G) \to \{1, 3\}$ by constructing a new edge coloring $c : E(G) \to \{1, 2, 3\}$ as follows:

$$c(e) = \begin{cases} \overline{c}(e) + 1 & \text{if } e = u_{\ell k + 1} u_{\lceil \frac{n}{2} \rceil} \text{ or} \\ & \quad e = u_{(j-1)k+1} u_{jk}, j = \ell - i, i \text{ odd}, 1 \leq i \leq \ell - \\ \overline{c}(e) - 1 & \text{if } e = u_{(j-1)k+1} u_{jk}, j = \ell + i, i \text{ odd}, 3 \leq i \leq \ell + 1 \\ \overline{c}(e) & \text{otherwise.} \end{cases}$$

Then the induced vertex coloring $c' : V(G) \to \mathbb{N}$ satisfies

$$c'(v) = \begin{cases} \overline{c}'(v) + 1 & \text{if } v = u_{\ell k + 1}, v = u_{\lceil \frac{n}{2} \rceil} \text{ or} \\ & \quad v = u_{(j-1)k+1}, u_{jk}, j = \ell - i, i \text{ odd}, 1 \leq i \leq \ell - 1 \\ \overline{c}'(v) - 1 & \text{if } v = u_{(j-1)k+1}, u_{jk}, j = \ell + i, i \text{ odd}, 3 \leq i \leq \ell + 1 \\ \overline{c}'(v) & \text{otherwise.} \end{cases}$$

It then follows by properties (1)–(3) of the vertex coloring $\bar{c}'$ that c' is vertex-distinguishing. This is illustrated for $K_{4(3)}$, $K_{5(3)}$ and $K_{4(5)}$ in the following three tables. □

$V(K_{4(3)})$	u_1	u_2	u_3	u_4	u_5	u_6	u_7	u_8	u_9	u_{10}	u_{11}	u_{12}
$\deg_H v$	9	8	7	6	6	5	5	4	3	2	1	0
$\bar{c}'(v)$	9	11	13	15	15	17	17	19	21	23	25	27
$c'(v)$	9	11	13	15	**16**	**18**	17	19	21	23	25	27

$V(K_{5(3)})$	u_1	u_2	u_3	u_4	u_5	u_6	u_7	u_8	u_9	u_{10}	u_{11}	u_{12}	u_{13}	u_{14}	u_{15}
$\deg_H v$	12	11	10	9	8	8	7	6	6	5	4	3	2	1	0
$\bar{c}'(v)$	12	14	16	18	20	20	22	24	24	26	28	30	32	34	36
$c'(v)$	12	14	16	18	20	**21**	22	**25**	24	26	28	30	32	34	36

$V(K_{4(5)})$	u_1, u_2, u_3, u_4	u_5, u_6, u_7, u_8	$u_9, u_{10}, u_{11}, u_{12}$	$u_{13}, u_{14}, u_{15}, u_{16}$	$u_{17}, u_{18}, u_{19}, u_{20}$
$\deg_H v$	15, 14, 13, 12	12, 11, 10, 9	9, 8, 8, 7	6, 5, 4, 3	3, 2, 1, 0
$\bar{c}'(v)$	15, 17, 19, 21	21, 23, 25, 27	27, 29, 29, 31	33, 35, 37, 39	39, 41, 43, 45
$c'(v)$	**16**, 17, 19, **22**	21, 23, 25, 27	**28**, **30**, 29, 31	33, 35, 37, 39	**38**, 41, 43, **44**

The following corollary then summarizes all results on the irregularity strength of regular complete multipartite graphs.

Corollary 2.11. *If G is a regular complete multipartite graph of order at least 3, then*

$$s(G) = \begin{cases} 4 & \text{if } G = K_{r,r} \text{ where } r \geq 3 \text{ is odd} \\ 3 & \text{otherwise.} \end{cases}$$

2.3 The Irregularity Strength of Paths and Cycles

We now turn our attention to two other well-known classes of graphs, namely paths and cycles. The next theorem gives the irregularity strength of all paths.

Theorem 2.12 ([24]). *For an integer $n \geq 3$,*

$$s(P_n) = \begin{cases} \frac{n}{2} & \text{if } n \equiv 0 \pmod 4 \\ \frac{n+1}{2} & \text{if } n \text{ is odd} \\ \frac{n+2}{2} & \text{if } n \equiv 2 \pmod 4. \end{cases}$$

Proof. Let $P_n = (v_1, v_2, \ldots, v_n)$ and $e_i = v_i v_{i+1}$ for $1 \leq i \leq n - 1$. First, we establish a lower bound for $s(P_n)$. If $c : E(P_n) \rightarrow \mathbb{N}$ is a vertex-distinguishing edge coloring with induced vertex coloring c', then $c'(v_j) \geq n$ for some vertex v_j. If v_j is an end-vertex, say $v_j = v_1$, then $c(e_1) \geq n$; while if $\deg v_j = 2$, then either $c(e_{j-1}) \geq n/2$ or $c(e_j) \geq n/2$. Thus $s(P_n) \geq n/2$ when n is even and $s(P_n) \geq (n+1)/2$ when n is odd. If $n \equiv 2 \pmod 4$ and $s(P_n) = n/2$, then $\{c'(v_i) : 1 \leq i \leq n\} = [n]$ and so $\sum_{i=1}^{n} c'(v_i)$ is odd, contradicting Proposition 2.1. Hence $s(P_n) \geq (n+2)/2$ when $n \equiv 2 \pmod 4$.

Next, we show that each of these lower bounds for $s(P_n)$ is also an upper bound. If $n \equiv 0 \pmod 4$, then $n = 4k$ for some positive integer k. Define the edge coloring $c : E(P_n) \to \mathbb{N}$ by

$$c(e_i) = \begin{cases} i & \text{if } 1 \leq i \leq 2k \\ n - 2\left\lfloor \frac{i}{2} \right\rfloor & \text{if } 2k + 1 \leq i \leq n - 1. \end{cases}$$

For the induced vertex coloring c', we then have

$$c'(v_i) = \begin{cases} 2i - 1 & \text{if } 1 \leq i \leq 2k \\ 2n - 2i + 2 & \text{if } 2k + 1 \leq i \leq n. \end{cases}$$

This is illustrated in Fig. 2.6 for $n = 8$. Since c is a vertex-distinguishing edge coloring whose largest color is $c(e_{2k}) = 2k = n/2$, it follows that $s(P_n) \leq n/2$ and so $s(P_n) = n/2$ if $n \equiv 0 \pmod 4$.

Assume next that n is odd. Then $n = 2k + 1$ for some positive integer k. If $n \equiv 3 \pmod 4$, then define the edge coloring $c : E(P_n) \to \mathbb{N}$ by

$$c(e_i) = \begin{cases} i & \text{if } 1 \leq i \leq k \\ n + 1 - 2\left\lceil \frac{i}{2} \right\rceil & \text{if } k + 1 \leq i \leq n - 1. \end{cases}$$

Then the induced vertex coloring c' is given by

$$c'(v_i) = \begin{cases} 2i - 1 & \text{if } 1 \leq i \leq k + 1 \\ 2n - 2i + 2 & \text{if } k + 2 \leq i \leq n. \end{cases}$$

If $n \equiv 1 \pmod 4$, then define the edge coloring $c : E(P_n) \to \mathbb{N}$ by

$$c(e_i) = \begin{cases} i & \text{if } 1 \leq i \leq k - 1 \text{ or } i = k + 1 \\ k + 1 & \text{if } i = k \\ n + 1 - 2\left\lceil \frac{i}{2} \right\rceil & \text{if } k + 2 \leq i \leq n - 1. \end{cases}$$

Then the induced vertex coloring c' is given by

$$c'(v_i) = \begin{cases} 2i - 1 & \text{if } 1 \leq i \leq k - 1 \\ 2i & \text{if } i = k, k + 1 \\ 2i - 3 & \text{if } i = k + 2 \\ 2n - 2i + 2 & \text{if } k + 3 \leq i \leq n. \end{cases}$$

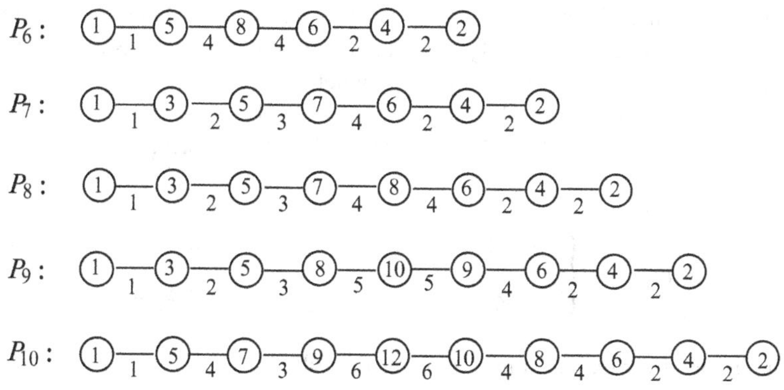

$P_6:$ ①—1—⑤—4—⑧—4—⑥—2—④—2—②

$P_7:$ ①—1—③—2—⑤—3—⑦—4—⑥—2—④—2—②

$P_8:$ ①—1—③—2—⑤—3—⑦—4—⑧—4—⑥—2—④—2—②

$P_9:$ ①—1—③—2—⑤—3—⑧—5—⑩—5—⑨—4—⑥—2—④—2—②

$P_{10}:$ ①—1—⑤—4—⑦—3—⑨—6—⑫—6—⑩—4—⑧—4—⑥—2—④—2—②

Fig. 2.6 Edge colorings of P_n in the proof of Theorem 2.12 for $6 \le n \le 10$

This is illustrated in Fig. 2.6 for $n = 7, 9$. In each case, c is a vertex-distinguishing edge coloring whose largest color is $c(e_{k+1}) = (n + 1)/2$, it follows that $s(P_n) \le (n + 1)/2$ and so $s(P_n) = (n + 1)/2$ when n is odd.

Finally, assume that $n \equiv 2 \pmod 4$. Then $n = 4k+2$ for some positive integer k. Define the edge coloring $c : E(P_n) \to \mathbb{N}$ by

$$c(e_i) = \begin{cases} i & \text{if } i = 1, 3 \cdots, 2k - 1 \\ i + 2 & \text{if } i = 2, 4 \cdots, 2k \\ n - 2 \lfloor \frac{i}{2} \rfloor & \text{if } 2k + 1 \le i \le n - 1. \end{cases}$$

Then the induced vertex coloring c' is given by

$$c'(v_i) = \begin{cases} 1 & \text{if } i = 1 \\ 2i + 1 & \text{if } 2 \le i \le 2k \\ 2n - 2i + 2 & \text{if } 2k + 1 \le i \le n. \end{cases}$$

This is illustrated in Fig. 2.6 for $n = 6, 10$. Since c is a vertex-distinguishing edge coloring having the largest color $c(e_{2k+1}) = (n + 2)/2$, it follows that $s(P_n) \le (n + 2)/2$ and so $s(P_n) = (n + 2)/2$ when $n \equiv 2 \pmod 4$. □

The next theorem gives the irregularity strength of cycles (see [41]).

Theorem 2.13. *For an integer $n \ge 3$,*

$$s(C_n) = \begin{cases} \frac{n+1}{2} & \text{if } n \equiv 1 \pmod 4 \\ \frac{n+2}{2} & \text{if } n \text{ is even} \\ \frac{n+3}{2} & \text{if } n \equiv 3 \pmod 4. \end{cases} \tag{2.11}$$

Proof. By Corollaries 2.6 and 2.7, each of the expressions in (2.11) is a lower bound for $s(C_n)$. Thus, it remains to verify that each of these expressions is also an upper bound. Let $C_n = (v_1, v_2, \ldots, v_n, v_{n+1} = v_1)$ where $n \geq 3$.

We first consider the case when $n \equiv 1 \pmod 4$. Then $n = 4q + 1$ for some positive integer q and so $\frac{n+1}{2} = 2q + 1$. Define an edge coloring $c : E(C_{4q+1}) \to [2q + 1]$ by

$$c(v_i v_{i+1}) = \begin{cases} 2q + 1 - 2\left\lfloor \frac{i}{2} \right\rfloor & \text{for } 1 \leq i \leq 2q + 1 \\ i - 2q & \text{for } 2q + 2 \leq i \leq 4q + 1. \end{cases}$$

Then the induced vertex coloring c' satisfies the following

$$c'(v_i) = \begin{cases} 4q + 4 - 2i & \text{if } 1 \leq i \leq 2q + 1 \\ 2i - 4q - 1 & \text{if } 2q + 2 \leq i \leq 4q + 1. \end{cases}$$

This is illustrated in Fig. 2.7 for C_9 and C_{13}. Since c is a vertex-distinguishing edge coloring whose largest color is $2q + 1$, it follows that $s(C_{4q+1}) \leq 2q + 1$ and so $s(C_{4q+1}) = 2q + 1$.

Next, we show that if n is even, the lower bound $\frac{n+2}{2}$ for $s(C_n)$ is also an upper bound. Then $n = 2k$ for some integer $k \geq 2$ and so $\frac{n+2}{2} = k + 1$. Define an edge coloring $c : E(C_{2k}) \to [k + 1]$ by considering two cases, according to whether k is odd or k is even. If k is odd, then let

$$c(v_i v_{i+1}) = \begin{cases} k + 1 - 2\left\lfloor \frac{i}{2} \right\rfloor & \text{for } 1 \leq i \leq k \\ k + 2 - 2\left\lfloor \frac{i}{2} \right\rfloor = 1 & \text{for } i = k + 1, k + 2 \\ i + 1 - k & \text{for } k + 3 \leq i \leq 2k \end{cases}$$

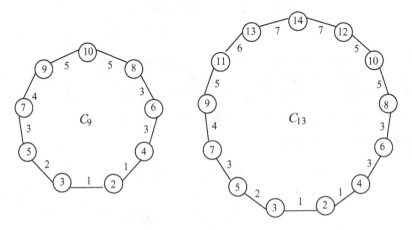

Fig. 2.7 Edge colorings of C_9 and C_{13}

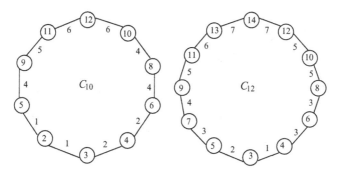

Fig. 2.8 Edge colorings of C_{10} and C_{12}

Then the induced vertex coloring c' satisfies the following

$$c'(v_i) = \begin{cases} 2k+4-2i & \text{if } 1 \leq i \leq k \\ 3 & \text{if } i = k+1 \\ 2 & \text{if } i = k+2 \\ 5 & \text{if } i = k+3 \\ 2i-2k+1 & \text{if } k+4 \leq i \leq 2k. \end{cases}$$

This is illustrated in Fig. 2.8 for C_{10}. If k is even, then let

$$c(v_i v_{i+1}) = \begin{cases} k+1-2\lfloor \frac{i}{2} \rfloor & \text{for } 1 \leq i \leq k \\ i+1-k & \text{for } k+1 \leq i \leq 2k \end{cases}$$

Then the induced vertex coloring c' satisfies the following

$$c'(v_i) = \begin{cases} 2k+4-2i & \text{if } 1 \leq i \leq k \\ 2i-2k+1 & \text{if } k+1 \leq i \leq 2k. \end{cases}$$

This is illustrated in Fig. 2.8 for C_{12}. Since c is a vertex-distinguishing edge coloring whose largest color is $k+1$, it follows that $s(C_{2k}) \leq k+1$ and so $s(C_{2k}) = k+1$.

Finally, we consider the case where n is odd and $n \equiv 3 \pmod 4$. In this case, $n = 4q+3$ for some positive integer q. Define an edge coloring $c : E(C_{4q+3}) \rightarrow [2q+3]$ by

$$c(v_i v_{i+1}) = \begin{cases} 2q+3-2\lfloor \frac{i}{2} \rfloor & \text{for } 1 \leq i \leq 2q+3 \\ i-(2q+2) & \text{for } 2q+4 \leq i \leq 4q+2 \\ 2q+3 & \text{for } i = 4q+3. \end{cases}$$

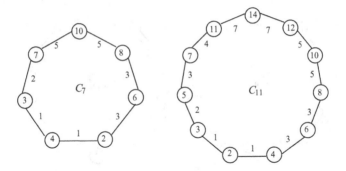

Fig. 2.9 Edge colorings of C_7 and C_{11}

Then the induced vertex coloring c' satisfies the following

$$c'(v_i) = \begin{cases} 4q + 8 - 2i & \text{if } 1 \leq i \leq 2q + 3 \\ 2i - 4q - 5 & \text{if } 2q + 4 \leq i \leq 4q + 2 \\ 4q + 3 & \text{if } i = 4q + 3. \end{cases}$$

This is illustrated in Fig. 2.9 for C_7 and C_{11}. Since c is a vertex-distinguishing edge coloring whose largest color is $2q + 3$, it follows that $s(C_{4q+3}) \leq 2q + 3$ and so $s(C_{4q+3}) = 2q + 3$.   □

2.4   Additional Bounds for the Irregularity Strength of a Graph

A graph G is said to be *factorable* into the factors (spanning subgraphs of G) $F_1, F_2, \ldots, F_t$ if these factors are (pairwise) edge-disjoint and $\cup_{i=1}^{t} E(F_i) = E(G)$. If G is factored into $F_1, F_2, \ldots, F_t$, then $\{F_1, F_2, \ldots, F_t\}$ is called a *factorization* of G. If a graph G has a factorization into two factors, one of which is regular, then the irregularity strength of the other factor provides an upper bound for the irregularity strength of G.

Proposition 2.14. *If $\{F_1, F_2\}$ is a factorization of a graph G where F_2 is regular, then $s(G) \leq s(F_1)$.*

Proof. Suppose that $s(F_1) = s$ and F_2 is r-regular. Then there is a vertex-distinguishing edge coloring $\bar{c} : E(F_1) \rightarrow \{1, 2, \ldots, s\}$. Let $c : E(G) \rightarrow \{1, 2, \ldots, s\}$ be the edge coloring where

$$c(e) = \begin{cases} \bar{c}(e) & \text{if } e \in E(F_1) \\ 1 & \text{if } e \in E(F_2). \end{cases}$$

Fig. 2.10 Illustrating that the inequality in Proposition 2.14 can be strict

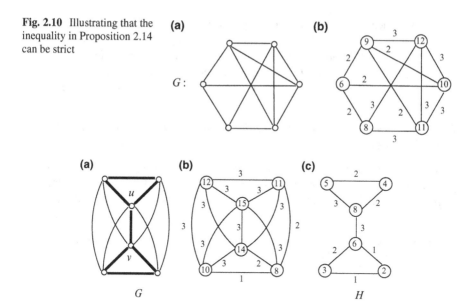

Fig. 2.11 Illustrating the equality in Proposition 2.14

Then $c'(v) = \overline{c}'(v) + r$ for every vertex v of G. Hence c is a vertex-distinguishing edge coloring of G and so $s(G) \leq s(F_1)$.   □

The inequality in Proposition 2.14 can be strict. For example, consider the graph G of Fig. 2.10a. Thus, $\{P_6, C_6\}$ is a factorization of G. By Theorem 2.12, $s(P_6) = \frac{6+2}{2} = 4$. The edge coloring of G in Fig. 2.10b shows that $s(G) \leq 3$. If there was an edge coloring of G with the colors in $\{1, 2\}$, then the vertex colors would belong to the set $\{3, 4, 5, 6, 7, 8\}$. However, there must be an even number of vertices of odd color by Proposition 2.1, a contradiction. Hence $s(G) \geq 3$, implying that $s(G) = 3$.

Equality in Proposition 2.14 is also possible. For example, consider the graph G of Fig. 2.11a with the spanning subgraph H. Thus, $\{H, C_6\}$ is a factorization of G, where the edges of H are drawn by bold lines in Fig. 2.11a. We show that $s(G) = s(H) = 3$. First, we verify that $s(G) = 3$. The edge coloring of G in Fig. 2.11b shows that $s(G) \leq 3$. If there was an edge coloring of G with the colors in $\{1, 2\}$, then the set of vertex colors would be a subset of $S = \{4, 5, 6, 7, 8, 9, 10\}$. So only one element in S is not a vertex color of G, necessarily an odd color. This implies that 10 is the color of some vertex of G. So either u or v has all of its incident edges colored 2, say u is colored 10. Since u is adjacent to all other vertices of G, every vertex is incident with an edge colored 2. Thus, the colors of the vertices of degree 4 are $2+1+1+1 = 5$, $2+2+1+1 = 6$, $2+2+2+1 = 7$ and $2+2+2+2 = 8$. Since 5 and 7 are two vertex colors, 9 is not a vertex color. Hence 4 must be the color of v. Since $\deg v = 5$, this is impossible. Hence $s(G) \geq 3$, implying that $s(G) = 3$.

Next, we show that $s(H) = 3$. If $s(H) = 2$, then for an edge coloring of H with the colors in $\{1, 2\}$, the set of vertex colors of H is a subset of $\{2, 3, 4, 5, 6\}$. This is impossible since the order of H is 6. Thus, $s(H) \geq 3$. Since the edge coloring of H with colors $1, 2, 3$ in Fig. 2.11c is vertex-distinguishing, it follows that $s(H) = 3$.

In 1952, Dirac [27] obtained the first theoretical result dealing with Hamiltonian graphs when he proved that if G is a graph of order $n \geq 3$ such that $\delta(G) \geq n/2$, then G is Hamiltonian. The following is a consequence of Dirac's theorem, Theorem 2.13 and Proposition 2.14.

Corollary 2.15 ([41]). *If G is an r-regular graph of order $n \geq 3$ such that $r \geq n/2$, then $s(G) \leq \left\lceil \frac{n}{2} \right\rceil + 1$.*

While the lower bounds for the irregularity strength of a graph G that we have presented thus far have been expressed primarily in terms of the order of G and the degrees of the vertices of G, the following lower bound is given in terms of the order and the size of G.

Theorem 2.16. *Let G be a connected graph of order $n \geq 3$ and size m. For each integer k with $2 \leq k \leq \Delta(G)$,*

$$s(G) \geq \frac{kn - 2m}{\binom{k}{2}}.$$

Proof. Let c be a vertex-distinguishing edge coloring of G with irregularity strength $s(G) = s$ where c' is the induced vertex coloring of G. For $i = 1, 2, \ldots, \Delta(G)$, let n_i denote the number of vertices of degree i in G. For $2 \leq k \leq \Delta(G)$,

$$1 \cdot n_1 + 2 \cdot n_2 + \cdots + (k-1)n_{k-1} + k(n - n_1 - n_2 - \cdots - n_{k-1}) = \sum_{i=1}^{\Delta(G)} i n_i = 2m.$$

Hence,

$$kn = (k-1)n_1 + (k-2)n_2 + \cdots + 1 \cdot n_{k-1} + 2m = \sum_{j=1}^{k-1}\left(\sum_{i=1}^{j} i n_i\right) + 2m.$$

For each integer j ($1 \leq j \leq k-1$), the colors of the $\sum_{i=1}^{j} i n_i$ vertices lie between 1 and js. Therefore,

$$\sum_{i=1}^{j} i n_i \leq js.$$

Thus,

$$kn = \sum_{j=1}^{k-1}\left(\sum_{i=1}^{j} i n_i\right) + 2m \leq \left(\sum_{j=1}^{k-1}(js)\right) = s\binom{k}{2} + 2m$$

and so

$$s = s(G) \geq \frac{kn - 2m}{\binom{k}{2}},$$

as desired. □

For $k = 2$ in Theorem 2.16, we have $s(G) \geq 2n - 2m$. Since G is connected, $m \geq n - 1$. If $m = n - 1$ (and so G is a tree), then $s(G) \geq 2$, which, of course, we already knew. If $k = 3$ in Theorem 2.16, we have $s(G) \geq \frac{3n-2m}{3}$, while if $k = 4$ in Theorem 2.16, we have $s(G) \geq \frac{4n-2m}{6} = \frac{2n-m}{3}$. When G is a tree, $\frac{3n-2m}{3} = \frac{n+2}{3}$ is a better bound for $s(G)$.

Corollary 2.17 ([24]). *If T is a tree of order $n \geq 3$, then $s(T) \geq (n + 2)/3$.*

Following [24], we now see that the lower bound $(n + 2)/3$ for $s(T)$, where T is a tree of order $n \geq 3$, cannot be improved in general by providing an infinite class of trees T of order $n \geq 3$ for which $s(T) = (n + 2)/3$.

For a positive integer q, let $P_{4q-1} = (u_1, u_2, \ldots, u_{4q-1})$ be a path of order $4q - 1$. We attach two paths (u_1, v_1, w_1) and (u_1, v_2, w_2) of length 2 at u_1. For $2 \leq i \leq 4q - 2$, attach a path (u_i, v_{i+1}, w_{i+1}) of length 2 at u_i. In addition, attach two paths $(u_{4q-1}, v_{4q}, w_{4q})$ and $(u_{4q-1}, v_{4q+1}, w_{4q+1})$ of length 2 at u_{4q-1}. Denote the resulting tree by T_q, which has order $12q + 1$. The tree T_2 is shown in Fig. 2.12. By Corollary 2.17, $s(T_q) \geq (n + 2)/3 = (12q + 3)/3 = 4q + 1$. It remains to show that $s(T_q) \leq 4q + 1$.

Define an edge coloring $c : E(T_q) \rightarrow [4q + 1]$ by $c(v_i w_i) = i$ for $1 \leq i \leq 4q + 1$, $c(u_1 v_1) = c(u_{4q-1} v_{4q+1}) = 4q + 1$ and $c(u_i v_{i+1}) = 4q + 1$ for $1 \leq i \leq 4q - 1$. In addition,

$$c(u_i u_{i+1}) = \begin{cases} 2q + \lceil \frac{i}{2} \rceil & \text{for } 1 \leq i \leq 2q + 1 \\ 2q + 1 + \lfloor \frac{i}{2} \rfloor & \text{for } 2q + 2 \leq i \leq 4q - 2. \end{cases}$$

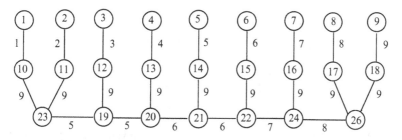

Fig. 2.12 An edge coloring of the tree T_2

This edge coloring c then induces the vertex coloring $c' : V(T_q) \to \mathbb{N}$ where

$$c'(w_i) = i \qquad \text{for } 1 \le i \le 4q + 1$$

$$c'(v_i) = 4q + 1 + i \quad \text{for } 1 \le i \le 4q + 1$$

$$c'(u_i) = \begin{cases} 10q + 3 & \text{if } i = 1 \\ 8q + 1 + i & \text{if } 2 \le i \le 2q + 1 \\ 8q + 2 + i & \text{if } 2q + 2 \le i \le 4q - 2 \\ 12q + 2 & \text{if } i = 4q - 1. \end{cases}$$

This is illustrated in Fig. 2.12 for T_2. Since c is a vertex-distinguishing edge coloring whose largest color is $4q + 1$, it follows that $s(T_q) \le 4q + 1$ and so $s(T_q) = 4q + 1$.

If G is a unicyclic graph of order n and size m, then $m = n$. Letting $k = 3$ in Theorem 2.16, we have the following corollary.

Corollary 2.18 ([24]). *If G is a unicyclic graph of order $n \ge 3$, then $s(G) \ge n/3$.*

Next, we show that the lower bound $n/3$ for $s(G)$, where G is a unicyclic graph of order $n \ge 3$, is sharp by providing an infinite class of unicyclic graphs G of order $n \ge 3$ for which $s(G) = n/3$.

For a positive integer q, let $C = (u_1, u_2, \ldots, u_{4q}, u_{4q+1} = u_1)$ be a cycle of length $4q$. At each vertex u_i ($1 \le i \le 4q$), we attach a path (u_i, v_i, w_i) of length 2. The resulting graph is a unicyclic graph G of order $12q$. By Corollary 2.18, $s(G) \ge 12q/3 = 4q$. To show that $s(G) \le 4q$, define an edge coloring $c : E(G) \to [4q]$ by

$$c(v_i w_i) = i \qquad \text{for } 1 \le i \le 4q$$

$$c(u_i v_i) = 4q \qquad \text{for } 1 \le i \le 4q$$

$$c(u_i u_{i+1}) = \begin{cases} 4q - \lfloor \frac{i-1}{2} \rfloor & \text{for } 1 \le i \le 2q + 1 \\ 4q - \lfloor \frac{i}{2} \rfloor & \text{for } 2q + 2 \le i \le 4q. \end{cases}$$

Then the induced vertex coloring c' satisfies the following

$$c'(w_i) = i \qquad \text{for } 1 \le i \le 4q$$

$$c'(v_i) = 4q + i \quad \text{for } 1 \le i \le 4q$$

$$c'(u_i) = \begin{cases} 10q & \text{if } i = 1 \\ 12q + 2 - i & \text{if } 2 \le i \le 2q + 1 \\ 12q + 1 - i & \text{if } 2q + 2 \le i \le 4q. \end{cases}$$

For $q = 2$, this graph is shown in Fig. 2.13. Since c is a vertex-distinguishing edge coloring whose largest color is $4q$, it follows that $s(G) \le 4q$ and so $s(G) = 4q$.

If G is a connected graphs of order n and size $n + 1$, then letting $k = 4$ in Theorem 2.16 provides a lower bound for $s(G)$.

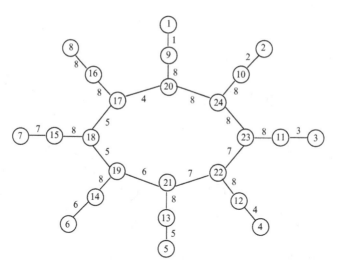

Fig. 2.13 An edge coloring of a unicyclic graph G

Corollary 2.19. *If G is a connected graph of order $n \geq 3$ and size $n + 1$, then*

$$s(G) \geq (n-1)/3.$$

Here too, we show that the lower bound $(n-1)/3$ for the irregularity strength of a connected graph of order $n \geq 3$ and size $n+1$, is sharp by providing an infinite class of connected graphs G of order $n \geq 3$ and size $n + 1$ for which $s(G) = (n-1)/3$.

For a positive integer q, let H be the graph of order $4q + 1$ and size $4q + 2$ consisting of two $(2q + 1)$-cycles

$$(x, u_1, u_2, \ldots, u_{2q}, x) \quad \text{and} \quad (x, u_{2q+1}, u_{2q+2}, \ldots, u_{4q}, x).$$

At each vertex u_i ($1 \leq i \leq 4q$) of H, we attach a path (u_i, v_i, w_i) of length 2. The resulting graph is a connected graph G of order $n = 12q + 1$ and size $12q + 2$. The graph G is shown in Fig. 2.14 for $q = 3$. By Corollary 2.19,

$$s(G) \geq (n-1)/3 = 12q/3 = 4q.$$

It remains to show that $s(G) \leq 4q$.

Define an edge coloring $c : E(G) \to [4q]$ by

$$c(v_i w_i) = i \qquad \text{for } 1 \leq i \leq 4q$$

$$c(u_i v_i) = 4q \qquad \text{for } 1 \leq i \leq 4q$$

$$c(x u_i) = 4q \qquad \text{for } i = 1, 4q$$

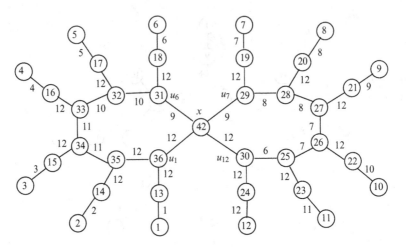

Fig. 2.14 An edge coloring of a connected graph of size $n + 1$

$$c(xu_i) = 3q \quad \text{for } i = 2q, 2q + 1$$

$$c(u_i u_{i+1}) = \begin{cases} 4q - \lfloor \frac{i}{2} \rfloor & \text{for } 1 \le i \le 2q - 1 \\ 4q - \lceil \frac{i}{2} \rceil & \text{for } 2q + 1 \le i \le 4q - 1. \end{cases}$$

Then the induced vertex coloring c' satisfies the following

$$c'(w_i) = i \qquad \text{for } 1 \le i \le 4q$$

$$c'(v_i) = 4q + i \quad \text{for } 1 \le i \le 4q$$

$$c'(u_i) = \begin{cases} 12q + 1 - i & \text{if } 1 \le i \le 2q \\ 12q - i & \text{if } 2q + 1 \le i \le 4q - 1. \\ 10q & \text{if } i = 4q \end{cases}$$

$$c'(x) = 14q.$$

This is illustrated in Fig. 2.14 for $q = 3$. Since c is a vertex-distinguishing edge coloring whose largest color is $4q$, it follows that $s(G) \le 4q$ and so $s(G) = 4q$.

While the results presented on irregularity strength have either dealt with formulas for the irregularity strength of certain classes of graphs or lower bounds, we now present a number of upper bounds. Since the proofs for these results are lengthy and do not provide additional insight into this topic, such results will be stated without proofs.

Theorem 2.20 ([3]). *If G is a connected graph of order $n \geq 4$, then $s(G) \leq n - 1$.*

Since a connected graph G of order $n \geq 3$ and size m has irregularity strength m if and only if G is a star and $m = n - 1$ in this case, the upper bound in Theorem 2.20 is sharp. Because the star of order n is the only tree whose irregularity strength is $n - 1$, there is an improved upper bound for other trees.

Theorem 2.21 ([3]). *If T is a tree of order $n \geq 4$ that is not a star, then $s(T) \leq n - 2$.*

Over the years, many research papers have dealt with the irregularity strength of special classes of graphs. For example, the papers [38, 41, 42] deal with the irregularity strength of regular graphs and [6, 13] concern trees. The papers [25, 40] discuss the irregularity strength of dense graphs (those graphs of order n and size m for which m/n is large). The irregularity strength of circulants and grids has been studied in [11, 26], respectively. Graphs with irregularity strength 2 were studied in [39].

Chapter 3
Modular Sum-Defined Irregular Colorings

In the sum-defined vertex-distinguishing edge coloring described in Chap. 1, the colors are selected from sets of the type $[k] = \{1, 2, \ldots, k\}$ for an integer $k \geq 2$. There is also a sum-defined vertex-distinguishing edge coloring whose colors are chosen instead from the sets $\mathbb{Z}_k$ of integers modulo k for integers $k \geq 2$.

For a connected graph G of order $n \geq 3$ and an integer $k \geq n$, let $c : E(G) \to \mathbb{Z}_k$ be an unrestricted edge coloring (where then adjacent edges may be colored the same). The edge coloring c induces a vertex coloring $c' : V(G) \to \mathbb{Z}_k$ defined by

$$c'(v) = \sum_{e \in E_v} c(e),$$

where the sum is computed in $\mathbb{Z}_k$. As in Chap. 2, we are interested in the case where c' is vertex-distinguishing. Such an edge coloring c always exists for some integers k. The minimum k for which there exists such a vertex-distinguishing edge coloring of G has been referred to as the *modular edge-gracefulness* meg(G) of G. Thus, meg$(G) \geq n$ for every connected graph G of order $n \geq 3$. If meg$(G) = n$, then G is called a *modular edge-graceful graph* and a vertex-distinguishing edge coloring $c : E(G) \to \mathbb{Z}_n$ is called a *modular edge-graceful labeling* as well as a *modular edge-graceful coloring* of G (which will be discussed in Sect. 3.2). This concept was introduced in 1991 by Gnana Jothi [48] under the terminology of *line-graceful graphs* (also see [47]). The choice of the terminology "modular edge-graceful graph" comes from the much studied concept of "graceful graph". It is useful to recall this concept.

© Ping Zhang 2015
P. Zhang, *Color-Induced Graph Colorings*, SpringerBriefs in Mathematics,
DOI 10.1007/978-3-319-20394-2_3

3.1   Graceful Graphs

Among the early mathematicians with an interest in decompositions of complete graphs was Anton Kotzig. One of Kotzig's students was Alexander Rosa in 1966. Kotzig and Rosa attended the Theory of Graphs International Symposium in Rome. Rosa spoke on "On certain valuations of the vertices of a graph". Rosa [75] defined a *valuation* of a graph G of order n and size m as a one-to-one function $f : V(G) \to \mathbb{N} \cup \{0\}$ from the vertex set of G to the set of nonnegative integers that gives rise to values of its edges, where the value of an edge $e = uv$ is defined as $|f(u) - f(v)|$. By placing various conditions required of f and the values of the edges of G, four different valuations were introduced: α-valuations, β-valuations, γ-valuations, ρ-valuations. In particular, the *β-valuation* requires that

$$f : V(G) \to \{0, 1, 2, \ldots, m\}$$

and

$$\{|f(u) - f(v)| : uv \in E(G)\} = \{1, 2, \ldots, m\}.$$

For example, β-valuations of C_3 and C_4 are shown in Fig. 3.1.

Golomb [49] referred to a β-valuation of a graph as a *graceful labeling* and a graph admitting a graceful labeling as a *graceful graph*. It is this terminology that became standard. In fact, many of graph labelings that have been studied since then have been patterned after graceful labelings. Thus, C_3 and C_4 are graceful. The three graphs shown in Fig. 3.2 are the only connected graphs of order 5 that are not graceful.

The following conjecture is due to Anton Kotzig and Gerhard Ringel.

The Graceful Tree Conjecture   Every tree is graceful.

Fig. 3.1 Illustrating
β-valuations of C_3 and C_4

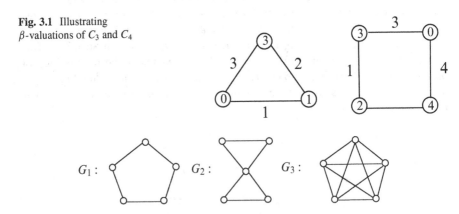

Fig. 3.2 Three graphs that are not graceful

We saw that $s(K_3) = 3$ by Theorem 2.4 and $s(P_3) = 2$ by Theorem 2.12. Furthermore, by Theorem 2.20 if G is a connected graph of order $n \geq 4$, then $s(G) \leq n - 1$. Since $\text{meg}(G) \geq n$, as mentioned above, we have the following observation.

Proposition 3.1. *If G is a connected graph of order at least 3 and $G \neq K_3$, then*

$$s(G) < \text{meg}(G).$$

3.2 Modular Edge-Graceful Graphs

First, we consider those connected graphs G of order n for which $\text{meg}(G) = n$ as well as the corresponding edge colorings, that is, modular edge-graceful graphs and modular edge-graceful colorings. To illustrate these concepts, consider the graphs G_1 and G_2 in Fig. 3.3. Since each of G_1 and G_2 has a modular edge-graceful coloring, as shown in Fig. 3.3, it follows that G_1 and G_2 are both modular edge-graceful. On the other hand, the graph $G_3 = C_6$ is not modular edge-graceful, that is, C_6 has order 6 but has no modular edge-graceful coloring. To see why C_6 fails to have a modular edge-graceful coloring, let $C_6 = (v_1, v_2, v_3, v_4, v_5, v_6, v_7 = v_1)$ where $e_i = v_i v_{i+1}$ for $1 \leq i \leq 6$ and suppose, to the contrary, that C_6 has a modular edge-graceful coloring, say $c : E(C_6) \to \mathbb{Z}_6$, where $c(e_i) \in \mathbb{Z}_6$. Then $c'(v_i) = c(e_i) + c(e_{i-1})$ (where $e_0 = e_6$). Since $\{c'(v_i) : 1 \leq i \leq 6\} = \mathbb{Z}_6$, it follows that

$$\sum_{i=1}^{6} c'(v_i) \equiv 3 \pmod 6.$$

However, $\sum_{i=1}^{6} c'(v_i) = 2 \sum_{i=1}^{6} c(e_i)$ and so

$$\sum_{i=1}^{6} c'(v_i) \equiv 0, 2 \text{ or } 4 \pmod 6,$$

which is a contradiction.

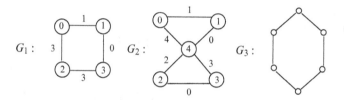

Fig. 3.3 Two modular edge-graceful graphs and a non-modular edge-graceful graph

The fact that C_6 is not modular edge-graceful illustrates a more general observation, which provides a necessary condition for a graph to be modular edge-graceful.

Proposition 3.2 ([48]). *If G is a modular edge-graceful connected graph of order* $n \geq 3$, *then* $n \not\equiv 2 \pmod 4$.

Proof. Assume, to the contrary, that there exists a modular edge-graceful graph of order $n \geq 3$ where $n \equiv 2 \pmod 4$. Let $c : E(G) \rightarrow \mathbb{Z}_n$ be a modular edge-graceful coloring of G and $c' : V(G) \rightarrow \mathbb{Z}_n$ the induced vertex coloring. Hence

$$\{c'(v) : v \in V(G)\} = \mathbb{Z}_n$$

and so

$$\sum_{v \in V(G)} c'(v) \equiv \frac{n}{2} \pmod n,$$

where $n/2$ is odd since $n \equiv 2 \pmod 4$. On the other hand, observe that

$$\sum_{v \in V(G)} c'(v) = 2 \sum_{uv \in E(G)} c(uv),$$

implying that $\sum_{v \in V(G)} c'(v)$ is even, a contradiction. $\qquad\square$

There is a useful observation concerning the two modular edge-graceful graphs in Fig. 3.3. One edge of G_1 is colored 0 and two edges of G_2 are colored 0. Of course, these edges contribute 0 to the color of any incident vertex. Thus, if we were to delete the edges colored 0 from G_1 and G_2, then we arrive at two graphs, trees in both cases (see Fig. 3.4), that also possess modular edge-graceful colorings. As we will see, the examples presented above illustrate more general observations.

Proposition 3.3. *If G is a modular edge-graceful connected graph, then every graph containing G as a spanning subgraph is also modular edge-graceful.*

Proof. Let H be a graph of order $n \geq 3$ containing a spanning subgraph G that is modular edge-graceful. Thus there exists a modular edge-graceful coloring $c_G : E(G) \rightarrow \mathbb{Z}_n$ of G such that the induced vertex coloring c'_G is vertex-distinguishing. The edge coloring c_G of G can then be extended to an edge coloring $c : E(H) \rightarrow \mathbb{Z}_n$ of H defined by $c(uv) = c_G(uv)$ if $uv \in E(G)$ and $c(uv) = 0$ if $uv \in E(H) - E(G)$. Since $c'(v) = c'_G(v)$ for each vertex v of H, it follows that c' is also vertex-distinguishing and so H is modular edge-graceful. $\qquad\square$

Fig. 3.4 Two modular edge-graceful trees

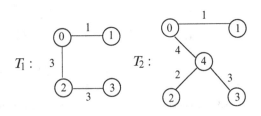

By Proposition 3.3, a connected graph is modular edge-graceful if it contains a modular edge-graceful spanning tree. This places additional importance on determining trees that are modular edge-graceful. It was conjectured by Gnana Jothi [48] that if T is a tree of order $n \geq 3$ for which $n \not\equiv 2 \pmod 4$, then T is modular edge-graceful (see [47]). This conjecture was verified in [59, 61] and so, by Proposition 3.3, the conjecture is not only true for trees but for all connected graphs. In order to present a proof of this fact, we first state some information that will be useful to us. It was verified in [64] that every tree of order $n \geq 3$ where $n \not\equiv 2 \pmod 4$ and diameter at least 5 is modular edge-graceful. This is stated formally below.

Theorem 3.4 ([64]). *A tree of order $n \geq 3$ having diameter at most 5 is modular edge-graceful if and only if $n \not\equiv 2 \pmod 4$.*

In order to present a result dealing with modular edge-graceful graphs that is along the same lines as a theorem by Bondy and Chvátal dealing with Hamiltonian graphs and closures (see [14]), we first present a lemma.

Lemma 3.5. *Let G be a connected graph of order at least 3 containing two nonadjacent vertices u and v that are connected by a path of odd length. Then the graph $G + uv$ is modular edge-graceful if and only if G is modular edge-graceful.*

Proof. Since G is a connected spanning subgraph of $G + uv$, it then follows by Proposition 3.3 that if G is modular edge-graceful, then so is $G + uv$. For the converse, assume that $G + uv$ is modular edge-graceful and let $c_1 : E(G+uv) \to \mathbb{Z}_n$ be a modular edge-graceful coloring of $G + uv$. Suppose that P is a $u - v$ path of odd length in G, say $P = (u = v_1, v_2, \ldots, v_p = v)$ where $p \geq 4$ is even. Now define the edge coloring $c_2 : V(G) \to \mathbb{Z}_n$ of G by

$$c_2(e) = \begin{cases} c_1(e) & \text{if } e \notin E(P) \\ c_1(e) + c_1(uv) & \text{if } e = v_i v_{i+1}, \ 1 \leq i \leq p-1 \text{ and } i \text{ is odd} \\ c_1(e) - c_1(uv) & \text{if } e = v_i v_{i+1}, \ 2 \leq i \leq p-2 \text{ and } i \text{ is even.} \end{cases}$$

Since $c_2'(x) = c_1'(x)$ in $\mathbb{Z}_n$ for all $x \in V(G)$, it follows that c_2 is a modular edge-graceful coloring of G. Thus G is a modular edge-graceful graph. $\square$

Let G be a connected graph of order at least 3 and let $\mathscr{P}$ be a partition of $V(G)$ into two or more independent sets. Define the *odd path closure of G with respect to $\mathscr{P}$*, denoted by $C_o(G, \mathscr{P})$ (or simply by $C_o(G)$ if the partition $\mathscr{P}$ under consideration is clear), as the graph obtained from G by recursively joining pairs of nonadjacent vertices that belong to different independent sets in $\mathscr{P}$ and that are connected by a path of odd length in G. Repeated application of Lemma 3.5 gives us the following result on modular edge-graceful graphs and odd path closures.

Proposition 3.6. *Let G be a connected graph of order at least 3, let $\mathscr{P}$ be a partition of $V(G)$ into two or more independent sets and let $C_o(G)$ be the odd path closure of G with respect to $\mathscr{P}$. Then $C_o(G)$ is modular edge-graceful if and only if G is modular edge-graceful.*

Next, we show that the odd path closure of a connected bipartite graph of order at least 3 with respect to given partite sets is a complete bipartite graph.

Lemma 3.7. *Let G be a connected bipartite graph with partite sets U and W where $|U| = r$ and $|W| = s$ and $r + s \geq 3$. Then the odd path closure $C_o(G)$ of G with respect to the partition $\{U, W\}$ is $K_{r,s}$.*

Proof. First, observe that $C_o(G)$ is a bipartite graph with partite sets U and W. If $C_o(G) \neq K_{r,s}$, then there exist vertices $u \in U$ and $w \in W$ for which $uw \notin E(C_o(G))$. Since $C_o(G)$ is bipartite,

$$U = \{v \in V(C_o(G)) : d_{C_o(G)}(u, v) \text{ is even}\}$$

$$W = \{v \in V(C_o(G)) : d_{C_o(G)}(u, v) \text{ is odd}\}.$$

Since $w \in W$, it follows that $d_{C_o(G)}(u, w)$ is odd. Thus $uw \in E(C_o(G))$, which is a contradiction. $\square$

For positive integers a and b, let $S_{a,b}$ be the *double star* (a tree of diameter 3) of order $a + b$ whose *central vertices* (non-end-vertices) have degrees a and b. By Theorem 3.4, every double star $S_{a,b}$ is modular edge-graceful if $a + b \not\equiv 2 \pmod 4$. We are now prepared to present the following theorem that classifies all modular edge-graceful trees.

Theorem 3.8. *Let T be a tree of order $n \geq 3$. Then T is modular edge-graceful if and only if $n \not\equiv 2 \pmod 4$.*

Proof. We have seen that if $n \equiv 2 \pmod 4$, then T is not modular edge-graceful. For the converse, assume that $n \not\equiv 2 \pmod 4$. Let U and W be the partite sets of T with $|U| = r$ and $|W| = s$. By Lemma 3.7, the odd path closure $C_o(G)$ of G with respect to the partition $\{U, W\}$ is $K_{r,s}$. By Proposition 3.6, it suffices to show that $G = K_{r,s}$ is modular edge-graceful. If $r = 1$ or $s = 1$, then $K_{r,s}$ is a star and so it is modular edge-graceful by Theorem 3.4. If $r \geq 2$ and $s \geq 2$, then the double star $S_{r,s}$ of order $r + s$ is a modular edge-graceful spanning subgraph of $K_{r,s}$. It then follows by Proposition 3.3 that $K_{r,s}$ is modular edge-graceful. Therefore, T is modular edge-graceful by Proposition 3.6. $\square$

The following is thus a consequence of Proposition 3.3 and Theorem 3.8.

Theorem 3.9 ([61]). *A connected graph of order $n \geq 3$ is modular edge-graceful if and only if $n \not\equiv 2 \pmod 4$.*

3.3 Non-modular Edge-Graceful Graphs

We have seen for every connected graph G of order n that $\text{meg}(G) \geq n$ and that $\text{meg}(G) = n$ if and only if G is modular edge-graceful. Consequently, if G is not modular edge-graceful, then $\text{meg}(G) \geq n + 1$. Thus the modular edge-gracefulness of a graph G is a measure of how close G is to being modular edge-graceful. As an illustration, we determine the modular edge-gracefulness of stars of order $n \geq 6$ with $n \equiv 2 \pmod 4$.

Proposition 3.10 ([61]). *If $n \geq 6$ is an integer for which $n \equiv 2 \pmod 4$, then*

$$\text{meg}(K_{1,n-1}) = n + 1.$$

Proof. Let $G = K_{1,n-1}$ be a star whose central vertex v is adjacent to $v_1, v_2, \ldots, v_{n-1}$. Define a coloring $c : E(G) \rightarrow \mathbb{Z}_{n+1}$ by

$$c(vv_i) = \begin{cases} 0 & \text{if } i = 1 \\ -\frac{i}{2} & \text{if } i \text{ is even and } 2 \leq i \leq n - 1 \\ \frac{i+1}{2} & \text{if } i \text{ is odd and } 3 \leq i \leq n - 3 \\ \frac{n+2}{2} & \text{if } i = n - 1. \end{cases}$$

Thus $\{c(vv_i) : 1 \leq i \leq n - 1\} = \{0, -1, \pm 2, \pm 3, \ldots, \pm \frac{n-2}{2}, \frac{n+2}{2}\}$. Since $c'(v) = n/2$ and

$$c'(v_i) = \begin{cases} 0 & \text{if } i = 1 \\ -\frac{i}{2} & \text{if } i \text{ is even and } 2 \leq i \leq n - 1 \\ \frac{i+1}{2} & \text{if } i \text{ is odd and } 3 \leq i \leq n - 3 \\ \frac{n+2}{2} & \text{if } i = n - 1, \end{cases}$$

it follows that $c' : V(G) \rightarrow \mathbb{Z}_{n+1}$ is vertex-distinguishing and so c is a modular edge-graceful coloring and $\text{meg}(G) = n + 1$. $\qquad\square$

In fact, $\text{meg}(G)$ has been determined for every connected graph G that is not modular edge-graceful. In order to do this, we first present two lemmas.

Lemma 3.11. *If H is a connected spanning subgraph of a graph G of order at least 3, then $\text{meg}(G) \leq \text{meg}(H)$.*

Proof. Suppose that $\text{meg}(H) = k$. Let $c_H : E(H) \rightarrow \mathbb{Z}_k$ be an edge coloring of H such that the induced vertex coloring $c'_H : V(H) \rightarrow \mathbb{Z}_k$ is vertex-distinguishing. Define an edge coloring $c_G : E(G) \rightarrow \mathbb{Z}_k$ by $c_G(e) = c_H(e)$ if $e \in E(H)$ and $c_G(e) = 0$ if $e \in E(G) - E(H)$. Since the induced vertex coloring $c'_G : V(G) \rightarrow \mathbb{Z}_k$ has the property that $c'_G(v) = c'_H(v)$ for all $v \in V(G)$, it follows that c'_G is vertex-distinguishing. Thus $\text{meg}(G) \leq k = \text{meg}(H)$. $\qquad\square$

Lemma 3.12. *Let G be a connected graph of order at least 3, let $\mathscr{P}$ be a partition of $V(G)$ into two or more independent sets and let $C_o(G)$ be the odd path closure of G with respect to $\mathscr{P}$. Then $\mathrm{meg}(G) = \mathrm{meg}(C_o(G))$.*

Proof. Since G is a connected spanning subgraph of a graph $C_o(G)$, it follows that $\mathrm{meg}(G) \leq \mathrm{meg}(C_o(G))$ by Lemma 3.11. On the other hand, employing an argument similar to the proof of Lemma 3.5 shows that $\mathrm{meg}(C_o(G)) \leq \mathrm{meg}(G)$ and so $\mathrm{meg}(C_o(G)) = \mathrm{meg}(G)$. □

We now determine the modular edge-gracefulness of double stars that are not modular edge-graceful.

Theorem 3.13. *If G is a double star of order $n \geq 6$ with $n \equiv 2$ (mod 4), then*

$$\mathrm{meg}(G) = n + 1.$$

Proof. Suppose that G is a double star with central vertices u and v where u is adjacent to end-vertices $u_1, u_2, \ldots, u_r$ and v is adjacent to end-vertices $v_1, v_2, \ldots, v_s$. Thus $n = r + s + 2$ and so $r + s \equiv 0$ (mod 4). We consider two cases.

Case 1. Either $r \equiv 0$ (mod 4) and $s \equiv 0$ (mod 4) or $r \equiv 2$ (mod 4) and $s \equiv 2$ (mod 4). Define an edge coloring $c : E(G) \to \mathbb{Z}_{n+1}$ by

$$c(uu_i) = \begin{cases} 0 & \text{if } i = 1 \\ 1 & \text{if } i = 2 \\ \frac{i+1}{2} & \text{if } i \text{ is odd and } 3 \leq i \leq r-1 \\ -\frac{i}{2} & \text{if } i \text{ is even and } 4 \leq i \leq r \end{cases}$$

$$c(vv_i) = \begin{cases} \frac{r+i+1}{2} & \text{if } i \text{ is odd and } 1 \leq i \leq s-1 \\ -\frac{r+i}{2} & \text{if } i \text{ is even and } 2 \leq i \leq s \end{cases}$$

$$c(uv) = \frac{r+s+2}{2}.$$

Observe that

$$\{c(uu_i) : 1 \leq i \leq r\} = \left\{0, 1, \pm2, \pm3, \ldots, \pm\frac{r}{2}\right\} \text{ and}$$

$$\{c(vv_i) : 1 \leq i \leq s\} = \left\{\pm\frac{r+2}{2}, \pm\frac{r+4}{2}, \ldots, \pm\frac{r+s}{2}\right\}.$$

Hence, $\{c'(x) : x \in V(G)\} = \{0, 1, \pm2, \pm3, \ldots, \pm\frac{r+s}{2}, \frac{r+s}{2} + 1, \frac{r+s}{2} + 2\}$. Thus, the induced vertex coloring $c' : V(G) \to \mathbb{Z}_{n+1}$ is vertex-distinguishing.

Case 2. Either $r \equiv 1$ (mod 4) and $s \equiv 3$ (mod 4) or $r \equiv 3$ (mod 4) and $s \equiv 1$ (mod 4), say the former. Assume therefore that $r \equiv 1$ (mod 4) and $s \equiv 3$ (mod 4). Then $r \geq 1$ and $s \geq 3$. We consider two subcases, according to whether $r = 1$ or $r \geq 5$.

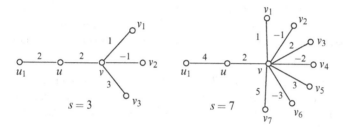

Fig. 3.5 The colorings in Subcase 2.1 for $s = 3$ and $s = 7$

Subcase 2.1. $r = 1$. Define an edge coloring $c : E(G) \to \mathbb{Z}_{n+1}$ by

$$c(uu_1) = \frac{1+s}{2}$$

$$c(vv_i) = \begin{cases} \frac{1+i}{2} & \text{if } i \text{ is odd and } 1 \leq i \leq s-2 \\ -\frac{i}{2} & \text{if } i \text{ is even and } 2 \leq i \leq s-1 \\ \frac{1+s}{2} + 1 & \text{if } i = s \end{cases}$$

$$c(uv) = 2.$$

Figure 3.5 shows the edge coloring c in the cases where $s = 3$ and $s = 7$. Observe that $\{c'(x) : x \in V(G)\} = \{\pm 1, \pm 2, \ldots, \pm \frac{s-1}{2}, \frac{s+1}{2}, \frac{s+1}{2} + 1\}$. Thus, the induced vertex coloring $c' : V(G) \to \mathbb{Z}_{n+1}$ is vertex-distinguishing.

Subcase 2.2. $r \geq 5$. Define an edge coloring $c : E(G) \to \mathbb{Z}_{n+1}$ by

$$c(uu_i) = \begin{cases} \frac{i+1}{2} & \text{if } i \text{ is odd and } 1 \leq i \leq r-2 \\ -\frac{i}{2} & \text{if } i \text{ is even and } 2 \leq i \leq r-1 \\ \frac{r+s}{2} & \text{if } i = r \end{cases}$$

$$c(vv_i) = \begin{cases} \frac{r+i}{2} & \text{if } i \text{ is odd and } 1 \leq i \leq s-2 \\ -\frac{r+i-1}{2} & \text{if } i \text{ is even and } 2 \leq i \leq s-1 \\ \frac{r+s+2}{2} & \text{if } i = s \end{cases}$$

$$c(uv) = 2.$$

Observe that

$$\{c(uu_i) : 1 \leq i \leq r\} = \left\{ \pm 1, \pm 2, \ldots, \pm \frac{r-1}{2}, \frac{r+s}{2} \right\}$$

$$\{c(vv_i) : 1 \leq i \leq s\} = \left\{ \pm \frac{r+1}{2}, \pm \frac{r+3}{2}, \ldots, \pm \frac{r+s-2}{2}, \frac{r+s+2}{2} \right\}.$$

Hence, $\{c'(x) : x \in V(G)\} = \{\pm 1, \pm 2, \ldots, \pm\frac{r+s-2}{2}, \frac{r+s}{2}, \frac{r+s}{2} + 1\}$. Thus, the induced vertex coloring $c' : V(G) \to \mathbb{Z}_{n+1}$ is vertex-distinguishing.

In each case, c is a modular edge-graceful coloring of G and so G is modular edge-graceful.   □

We have now seen that $\text{meg}(T) = n + 1$ for every star or double star T of order $n \geq 6$ when $n \equiv 2 \pmod 4$. With this information, we are prepared to show that $\text{meg}(T) = n + 1$ for every tree T that is not modular edge-graceful.

Theorem 3.14. *If T is a tree of order $n \geq 6$ with $n \equiv 2 \pmod 4$, then*

$$\text{meg}(T) = n + 1.$$

Proof. Suppose that the partite sets of T are U and W with $|U| = r$ and $|W| = s$. Then $n = r + s \equiv 2 \pmod 4$. By Lemma 3.7, the odd path closure $C_o(G)$ of G with respect to the partition $\{U, W\}$ is $K_{r,s}$. If $r = 1$ or $s = 1$, then $\text{meg}(K_{r,s}) = n + 1$ by Theorem 3.13. Thus, we may assume that $r \geq 2$ and $s \geq 2$. Then the double star $S_{r,s}$ is a spanning subgraph of $K_{r,s}$. Since $K_{r,s}$ is not modular edge-graceful, $\text{meg}(K_{r,s}) \geq n + 1$. On the other hand, $\text{meg}(S_{r,s}) = n + 1$ by Theorem 3.13. It then follows by Lemma 3.11 that $\text{meg}(K_{r,s}) \leq \text{meg}(S_{r,s}) = n + 1$ and so $\text{meg}(K_{r,s}) = n + 1$. Therefore, $\text{meg}(T) = n + 1$ by Lemma 3.12.   □

As a consequence of Lemma 3.12 and Theorem 3.14, we have the following result.

Theorem 3.15 ([61]). *For every connected graph G of order $n \geq 3$, either*

$$\text{meg}(G) = n \;\; or \;\; \text{meg}(G) = n + 1.$$

3.4 Nowhere-Zero Modular Edge-Graceful Graphs

If G is a modular edge-graceful connected spanning subgraph of a graph H, then each modular edge-graceful coloring of G can be extended to a modular edge-graceful coloring of H by assigning the color 0 to each edge of H that does not belong to G. Indeed, modular edge-graceful colorings of a graph that assign 0 to some edges of the graph played an important role in establishing Theorems 3.9 and 3.15. For this reason, it is of particular interest to consider modular edge-graceful colorings in which no edge is colored 0. This gives rise to a concept introduced in [60]. For a connected graph G of order $n \geq 3$, let $c : E(G) \to \mathbb{Z}_n - \{0\}$ where adjacent edges may be colored the same and let $c' : V(G) \to \mathbb{Z}_n$ be defined by $c'(v) = \sum_{e \in E_v} c(e)$, where the sum is computed in $\mathbb{Z}_n$. If c' is vertex-distinguishing, then c is called a *nowhere-zero modular edge-graceful labeling* or *nowhere-zero modular edge-graceful coloring*, also referred to as a *nowhere-zero meg-coloring*. If G has such an edge coloring, then G is called a *nowhere-zero modular edge-graceful graph* (see [60]). For example, consider the graph G of order 11 of Fig. 3.6 where two modular edge-graceful colorings of G are shown. In each coloring, an edge of G

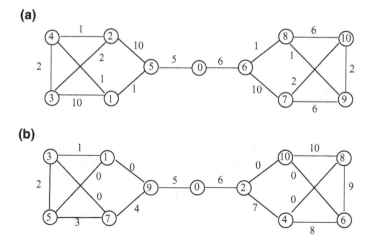

Fig. 3.6 Two modular edge-graceful colorings of a graph

is colored with an element in $\mathbb{Z}_{11} = \{0, 1, \ldots, 10\}$ and each vertex of G is colored with its induced color. The modular edge-graceful coloring of G in Fig. 3.6a is a nowhere-zero meg-coloring, while the modular edge-graceful coloring in Fig. 3.6b is not. Thus G is a nowhere-zero modular edge-graceful graph.

As an illustration, we determine all stars that are nowhere-zero modular edge-graceful.

Proposition 3.16 ([60]). *A star of order $n \geq 3$ is nowhere-zero modular edge-graceful if and only if n is odd.*

Proof. Let $G = K_{1,s}$ be a star of order $n = s + 1 \geq 3$ and let $V(G) = \{u, v_1, v_2, \ldots, v_s\}$ where u is the central vertex of G. We show that G is nowhere-zero modular edge-graceful if and only if s is even. First, we make an observation. If $c : E(G) \to \mathbb{Z}_n - \{0\}$ is a nowhere-zero modular edge-graceful coloring of G, then $c'(v_i) = c(uv_i) \neq 0$ for $1 \leq i \leq s$ and $c(uv_i) \neq c(uv_j)$ for all i, j with $1 \leq i \neq j \leq s$. Thus, we may assume that $c(uv_i) = i$ for $1 \leq i \leq s$ and so $c'(v_i) = i$. This implies that

$$c'(u) = 1 + 2 + \ldots + s = \binom{s+1}{2} = 0.$$

If s is even, then $c'(u) \equiv 0 \pmod{n}$; while if s is odd, then in $\mathbb{Z}_n$,

$$c'(u) = \binom{s+1}{2} = \frac{s(s+1)}{2} = \frac{(s+1)^2}{2} - \frac{s+1}{2} = n - \frac{s+1}{2} = \frac{s+1}{2}$$

and so $c'(u) \not\equiv 0 \pmod{n}$. Therefore, c is nowhere-zero modular edge-graceful if and only if s is even. □

The following result provides a characterization of connected nowhere-zero modular edge-graceful graphs.

Theorem 3.17 ([60]). *A connected graph G of order $n \geq 3$ is nowhere-zero modular edge-graceful if and only if*

1. $n \not\equiv 2 \pmod 4$,
2. $G \neq K_3$ *and*
3. *G is not a star of even order.*

For every connected graph G of order n, there is a smallest integer $k \geq n$ for which there exists an edge coloring $c : E(G) \to \mathbb{Z}_k - \{0\}$ such that the induced vertex coloring $c' : V(G) \to \mathbb{Z}_k$ defined by $c'(v) = \sum_{u \in N(v)} c(uv)$, where the sum is computed in $\mathbb{Z}_k$, is vertex-distinguishing. This number k is referred in [60] to as the *nowhere-zero modular edge-gracefulness* of G and is denoted by nzg(G). Thus nzg(G) $= n$ if and only if G is nowhere-zero modular edge-graceful and so nzg(G) $\geq n + 1$ if G is not nowhere-zero modular edge-graceful. For a connected graph G of order $n \geq 3$ with $n \not\equiv 2 \pmod 4$ that is not nowhere-zero modular edge-graceful, the exact value of nzg(G) has been determined.

Theorem 3.18 ([60]). *If G is a connected graph of order $n \geq 3$ with $n \not\equiv 2$ (mod 4) that is not nowhere-zero modular edge-graceful, then*

$$n + 1 \leq \text{nzg}(G) \leq n + 2.$$

Furthermore, nzg(G) $= n + 1$ if and only if $G = K_3$ and nzg(G) $= n + 2$ if and only if G is a star of even order.

By Theorem 3.9, if G is a connected graph of order $n \geq 6$ where $n \equiv 2 \pmod 4$, then G is not modular edge-graceful. Consequently, G is not nowhere-zero modular edge-graceful either and so nzg(G) $\geq n + 1$. For connected graphs of order $n \geq 3$ with $n \equiv 2 \pmod 4$, we then have the following result.

Theorem 3.19 ([60]). *If G is a connected graph of order $n \geq 6$ such that $n \equiv 2$ (mod 4), then $n + 1 \leq \text{nzg}(G) \leq n + 2$, where nzg($G$) $= n + 2$ if and only if G is a star.*

Chapter 4
Set-Defined Irregular Colorings

Chapter 2 dealt with vertex-distinguishing unrestricted edge colorings c of a graph G using the colors $1, 2, \ldots, k$ for some positive integer k that induce vertex colorings c' where $c'(v)$ is defined as the sum of the colors of the edges incident with v. Chapter 3 also explored vertex-distinguishing unrestricted edge colorings, where in that chapter the colors were members of some set $\mathbb{Z}_k$ of integers modulo k and the induced color of a vertex was once again the sum (computed in $\mathbb{Z}_k$) of the colors of edges incident with the vertex.

4.1 The Set Irregular Chromatic Index

Other vertex colorings have been defined from a given unrestricted edge coloring c. One of these was introduced by Harary and Plantholt [56], who defined a vertex coloring c' in which $c'(v)$ is the *set* of colors of the edges incident with vertex v. If c' is vertex-distinguishing, then c is called a *set irregular edge coloring*. The minimum positive integer k for which a graph G has a set irregular edge coloring is the *set irregular chromatic index* of G and is denoted by $\mathrm{si}(G)$. As with irregularity strength and modular edge-gracefulness, the set irregular chromatic index does not exist for K_2. Since every two vertices in a connected graph G of order $n \geq 3$ and size m are incident with different sets of edges, any edge coloring that assigns the edges of G distinct colors of $[m]$ is a set irregular edge coloring. Hence $\mathrm{si}(G)$ exists and $\mathrm{si}(G) \leq m$.

Whether one is considering an edge coloring that induces a sum-defined or a set-defined vertex coloring, the main requirement here is that such an edge coloring must be vertex-distinguishing and, preferably, accomplishes this using the minimum number of edge colors. We will see that defining vertex colorings by means of sets rather than sums ordinarily requires the use of more colors, however. For example, we saw that for every integer $n \geq 3$, the unique connected graph G_n of order n having

© Ping Zhang 2015
P. Zhang, *Color-Induced Graph Colorings*, SpringerBriefs in Mathematics,
DOI 10.1007/978-3-319-20394-2_4

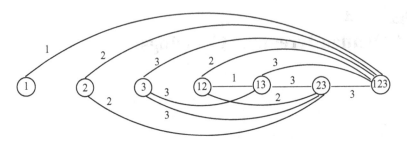

Fig. 4.1 The graph G_7 of order 7 with set irregular chromatic index 3

exactly two vertices of the same degree has irregularity strength 2. In particular, $s(G_7) = 2$. On the other hand, no set irregular edge coloring of G_7 exists if only the two colors 1 and 2 are used since the only nonempty subsets of $\{1, 2\}$ are $\{1\}, \{2\}$ and $\{1, 2\}$ but the order of G_7 is 7. Because the number of nonempty subsets of $\{1, 2, 3\}$ is $2^3 - 1 = 7$, it follows that $\mathrm{si}(G_7) \geq 3$. In fact, $\mathrm{si}(G_7) = 3$, as the edge coloring in Fig. 4.1 shows, where we denote the set $\{a\}$ by a, $\{a, b\}$ by ab and so on.

In general then, graphs of large order have large set irregular chromatic index.

Proposition 4.1. *If G is a connected graph of order $n \geq 3$, then*

$$\mathrm{si}(G) \geq \lceil \log_2(n + 1) \rceil.$$

Proof. Suppose that $\mathrm{si}(G) = k$ for some integer $k \geq 2$ and that c is a set irregular k-edge coloring of G using the colors in the set $[k] = \{1, 2, \ldots, k\}$. Since each induced color $c'(v)$ of a vertex v is a nonempty subset of $[k]$ and there are $2^k - 1$ nonempty subsets of $[k]$, it follows that $2^k - 1 \geq n$ and so $k = \mathrm{si}(G) \geq \lceil \log_2(n + 1) \rceil$. $\qquad\square$

We have seen that $s(G_n) = 2$ for the unique connected graph G_n of order $n \geq 3$ having exactly two vertices of the same degree. We show, for example, that if $n = 2^k - 1$ for some integer $k \geq 2$, then $\mathrm{si}(G_n) = k$.

Proposition 4.2. *For each integer $n = 2^k - 1$ where $k \geq 2$, $\mathrm{si}(G_n) = k$.*

Proof. By Proposition 4.1, $\mathrm{si}(G_n) \geq k$. It remains to show that $\mathrm{si}(G_n) \leq k$, that is, to show that there exists an edge coloring $c : E(G_n) \to [k]$ so that

$$\{c'(v) : v \in V(G_n)\} = \{A : A \subseteq [k], A \neq \emptyset\}. \tag{4.1}$$

List the n nonempty subsets of $[k]$ as $S_1, S_2, \ldots, S_n$ such that (a) $|S_i| \leq |S_j|$ for $1 \leq i < j \leq n$ and (b) $S_{n-i} = \bar{S}_i$ for every integer i with $1 \leq i < n/2$. In particular, $|S_i| = 1$ and $|S_{n-i}| = k - 1$ for $1 \leq i \leq k$. Also, $\left|S_{\frac{n+1}{2}}\right| = \left|S_{\frac{n-1}{2}}\right|$ if k is even and $\left|S_{\frac{n+1}{2}}\right| = \left|S_{\frac{n-1}{2}}\right| + 1$ if k is odd. For example, when $k = 3$ and $n = 2^k - 1 = 7$, a possible list of $S_1, S_2, \ldots, S_7$ of the seven nonempty subsets of $\{1, 2, 3\}$ is

$$\{1\}, \{2\}, \{3\}, \{1, 2\}, \{1, 3\}, \{2, 3\}, \{1, 2, 3\}.$$

Let $V(G_n) = \{v_1, v_2, \ldots, v_n\}$, where $\deg v_1 \leq \deg v_2 \leq \cdots \leq \deg v_n$. Thus, $v_i v_j \in E(G_n)$ if and only if $i + j \geq n + 1$. Furthermore, if i and j are positive integers such that $i + j \geq n + 1$, then $|S_i| + |S_j| \geq k$ and $S_i \cap S_j \neq \emptyset$. For an edge $v_i v_j$ of G_n, define $c(v_i v_j)$ to be any color belonging to $S_i \cap S_j$. For $1 \leq i \leq n$, let $\ell \in S_i$. Define $S_j = \overline{S_i} \cup \{\ell\}$. Then $S_i \cap S_j = \{\ell\}$, $|S_i| + |S_j| \geq k + 1$ and so $i + j \geq n + 1$. Hence $v_i v_j \in E(G_n)$ and $c(v_i v_j) = \ell$. Thus $\ell \in c'(v_i)$ and so $c'(v_i) = S_i$. Therefore, (4.1) holds and so $\mathrm{si}(G_n) \leq k$. Hence $\mathrm{si}(G_n) = k$. This edge coloring for $k = 3$ is shown in Fig. 4.1. $\qquad\square$

The following two results give the set irregular chromatic indices of two well-known classes of graphs.

Theorem 4.3 ([56]). *For every integer* $n \geq 3$,

$$\mathrm{si}(P_n) = \min\left\{ 2\left\lceil \frac{1 + \sqrt{8n - 9}}{4} \right\rceil - 1, \quad 2\left\lceil \sqrt{\frac{n-1}{2}} \right\rceil \right\}.$$

Theorem 4.4 ([56]). *For every integer* $n \geq 3$,

$$\mathrm{si}(C_n) = \min\left\{ 2\left\lceil \frac{1 + \sqrt{8n + 1}}{4} \right\rceil - 1, \quad 2\left\lceil \sqrt{\frac{n}{2}} \right\rceil \right\}.$$

4.2 Complete Graphs and Hypercubes

Next, we present formulas for the set irregular chromatic indices of two other common classes of graphs, beginning with complete graphs.

Theorem 4.5 ([56]). *For every integer* $n \geq 3$,

$$\mathrm{si}(K_n) = \lceil \log_2 n \rceil + 1.$$

Proof. Suppose for an integer $n \geq 3$ that there exists a set irregular k-edge coloring c of K_n using the colors in the set $[k] = \{1, 2, \ldots, k\}$, where the vertex v_i $(1 \leq i \leq n)$ in K_n is assigned the color subset S_i of $[k]$. Since v_i is adjacent to v_j for each integer $j \neq i$ where $1 \leq j \leq n$, it follows that $c(v_i v_j) \in S_i \cap S_j$ and so $S_i \cap S_j \neq \emptyset$. Consequently, $\overline{S_i}$ $(1 \leq i \leq n)$ is not a color for any vertex of K_n and so there are at most $\frac{1}{2}(2^k) = 2^{k-1}$ choices for the colors of the vertices of K_n. Therefore, $n \leq 2^{k-1}$ and so $\log_2 n \leq k - 1$. Thus $\mathrm{si}(K_n) \geq \lceil \log_2 n \rceil + 1$.

To show that $\mathrm{si}(K_n) \leq \lceil \log_2 n \rceil + 1$, it suffices to show that there is a set irregular edge coloring of K_n using colors in the set $\{1, 2, \ldots, \lceil \log_2 n \rceil + 1\}$. Figure 4.2 shows that such is the case for K_3 and K_4. Hence we may assume that $n \geq 5$.

Let $k = \lceil \log_2 n \rceil + 1$. Thus $2^{k-2} + 1 \leq n \leq 2^{k-1}$. Let $V(K_n) = \{v_0, v_1, v_2, \ldots, v_{n-1}\}$. First, we assign each vertex v_i $(0 \leq i \leq n - 1)$ a nonempty

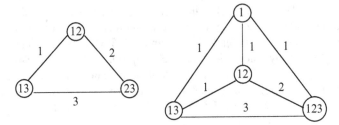

Fig. 4.2 Set irregular edge colorings of K_3 and K_4

Fig. 4.3 A set irregular
4-edge coloring of K_8

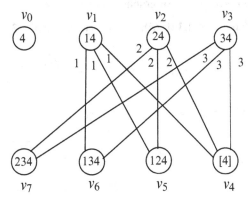

subset S_i of $[k]$. For $0 \leq i \leq k$, let $S_0 = \{k\}$, $S_i = \{i, k\}$ for $1 \leq i \leq k-1$ and $S_k = [k]$. For $k + 1 \leq i \leq n - 1$, choose the sets S_i in any manner so that $S_0, S_1, \ldots, S_{n-1}$ are distinct and $k \in S_i$. We next define an edge coloring c of K_n. For each integer i with $1 \leq i \leq k-1$, assign the color i to each edge $v_i v_t$ if $i \in S_t$ where $k \leq t \leq n-1$ and assign the color k to all other edges of K_n. Figure 4.3 shows such a 4-edge coloring of K_8, where edges not drawn are colored 4. Thus $c'(v_j) = S_j$ for all j $(0 \leq j \leq n - 1)$ and c is vertex-distinguishing. Therefore, $\text{si}(K_n) \leq k$ and so $\text{si}(G) = k = \lceil \log_2 n \rceil + 1$. $\qquad\square$

We next present a formula for the set irregular chromatic index of hypercubes.

Theorem 4.6 ([56]). *For every integer $n \geq 2$, $\text{si}(Q_n) = n + 1$.*

Proof. First, suppose that there is a set irregular edge coloring of Q_n using the colors in the set $[k]$. Since the order of Q_n is 2^n and there are $2^k - 1$ nonempty subsets of $[k]$, it follows that $2^n \leq 2^k - 1$ and so $k \geq n + 1$. Therefore, $\text{si}(Q_n) \geq n + 1$.

To show that $\text{si}(Q_n) \leq n + 1$, it suffices to show that there is a set irregular edge coloring of Q_n with the elements of $[n + 1]$. To accomplish this, we verify the following statement by induction on $n \geq 2$.

There is a set irregular edge coloring of Q_n using colors in the set $[n + 1]$ with the property that $V(Q_n)$ can be partitioned into two sets U and W of 2^{n-1} vertices each such that (i) each vertex in U is incident with one or more edges colored $n+1$ and (ii) there is a set S of edges colored 1 that are not adjacent to any edge colored $n + 1$ and each vertex in W is incident with at least one edge in S.

Fig. 4.4 A set irregular
3-edge coloring of $Q_2 = C_4$

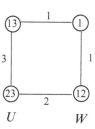

Figure 4.4 shows that this statement is true for $n = 2$, where the set S consists of the single edge colored 1 joining the two vertices in W.

Assume that an edge coloring exists for Q_k where $k \geq 2$ satisfying the statement above. We now consider Q_{k+1}. Since $Q_{k+1} = Q_k \square K_2$, we may assume that Q_{k+1} consists of two disjoint copies of Q_k, which we denote by Q'_k and Q''_k, and such that each pair of corresponding vertices in Q'_k and Q''_k is joined by an edge. We now describe a set irregular edge coloring of Q_{k+1} using the colors in $[k+2]$.

By the induction hypothesis, there is a set irregular edge coloring c_1 of Q'_k using the colors in $[k+1]$ with the prescribed properties (i) and (ii). Let c_2 be the edge coloring of Q''_k which is exactly the same as c_1 for Q'_k except that (1) the color $k+1$ in c_1 is replaced by $k+2$ and (2) the color 1 of any edge not adjacent to an edge colored $k+2$ is replaced by $k+1$. Thus c_2 is a set irregular edge coloring of Q''_k. In the subgraph Q''_k, there are exactly 2^{k-1} vertices that are incident with one or more edges colored $k+2$, there is a set of edges colored $k+1$ that are not adjacent to any edge colored $k+2$ and each of the remaining 2^{k-1} vertices is incident with at least one edge in this set. Define the edge coloring $c : E(Q_{k+1}) \rightarrow [k+2]$ by

$$c(e) = \begin{cases} c_1(e) & \text{if } e \in E(Q'_k) \\ c_2(e) & \text{if } e \in E(Q''_k) \\ k+2 & \text{if } e = u'u'' \text{ where } u' \in V(Q'_k),\ u'' \in V(Q''_k) \text{ and} \\ & \quad u'' \text{ is incident with an edge colored } k+2 \\ 1 & \text{otherwise.} \end{cases}$$

This is illustrated in Fig. 4.5 for Q_3 and Q_4.

Let U denote the set consisting of the 2^k vertices in Q_{k+1} that are incident with one or more edges colored $k+2$. Also, let

$$S = \left\{ u'u'' : c(u'u'') = 1, u' \in V(Q'_k) \text{ and } u'' \in V(Q''_k) \right\}.$$

Thus, no edge in S is adjacent to any edge colored $k+2$ and each vertex in $W = V(Q_{k+1}) - U$ is incident with an edge in S. Thus, c satisfies properties (i) and (ii) described in the statement. In the case of Q_3 shown in Fig. 4.5, the set U consists of the four solid vertices, $W = V(Q_3) - U$ and S consists of two edges colored 1 that join a vertex in Q'_2 and a vertex in Q''_2 (or the vertices in W).

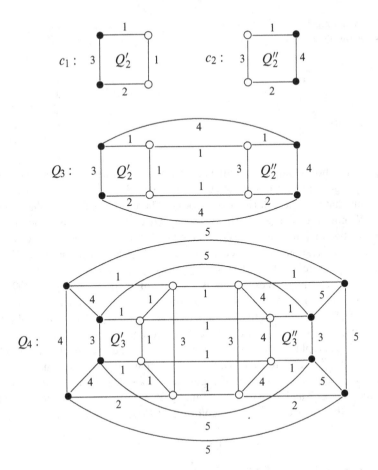

Fig. 4.5 Constructing a set irregular 4-edge coloring of Q_3 and a set irregular 5-edge coloring of Q_4

It remains to show that c is a set irregular edge coloring of Q_{k+1}. For each vertex x in Q_k', either $c'(x) = c_1'(x)$ or $c'(x) = c_1'(x) \cup \{k+2\}$. Since c_1 is a set irregular edge coloring of Q_k' using the colors in $[k+1]$, the colors $c'(x)$ of vertices x of Q_k' are distinct. For each vertex y in Q_k'', either

$$\text{(a) } c'(y) = c_2'(y) \text{ or (b) } c'(y) = c_2'(y) \cup \{1\} \neq c_2'(y).$$

Furthermore, (b) occurs only when $k+1 \in c_2'(y)$ and $k+2 \notin c_2'(y)$. On the other hand, if $1 \in c_2(y)$, then $k+1 \notin c_2'(y)$ and $k+2 \in c_2'(y)$. Since c_2 is a set irregular edge coloring of Q_k'', the colors $c'(y)$ of vertices y of Q_k'' are all distinct. Let $x \in V(Q_k')$ and $y \in V(Q_k'')$. If both $c'(x)$ and $c'(y)$ contain $k+2$, then $k+1 \in c'(x)$ and $k+1 \notin c'(y)$; while if neither $c'(x)$ nor $c'(y)$ contains $k+2$, then $k+1 \notin c'(x)$ and $k+1 \in c'(y)$. In either case, $c'(x) \neq c'(y)$. Thus c is vertex-distinguishing. The result then follows by the Principle of Mathematics Induction. □

4.3 Complete Bipartite Graphs

Figure 4.6 shows a set irregular 3-edge coloring of $K_{2,2}$, a set irregular 4-edge coloring of $K_{3,3}$ and a set irregular 4-edge coloring of $K_{4,4}$. Since the order of $K_{2,2}$ is $4 > 2^2 - 1$, it follows that $si(K_{2,2}) = 3$; while since the order of $K_{4,4}$ is $8 > 2^3 - 1$, it follows that $si(K_{4,4}) = 4$. Furthermore, $si(K_{3,3}) \le 4$. In fact, $si(K_{3,3}) = 4$, as we now verify. If there exists a set irregular 3-edge coloring of $K_{3,3}$, then exactly one of the seven nonempty subsets of $[3]$ is not used for the colors of the vertices of $K_{3,3}$ and so at least two of the sets $\{1\}$, $\{2\}$ and $\{3\}$ of $[3]$ are used, say $c'(u) = \{1\}$ and $c'(v) = \{2\}$ where $u, v \in V(K_{3,3})$. Necessarily, u and v belong to the same partite set of $K_{3,3}$. Hence the color of each vertex in the other partite set of $K_{3,3}$ must contain $\{1, 2\}$. However, there are only two such subsets, namely $\{1, 2\}$ and $\{1, 2, 3\}$, which is impossible. Therefore, $si(K_{3,3}) = 4$.

When $r \ge 5$, the value of $si(K_{r,r})$ is one of two numbers.

Theorem 4.7 ([56]). *For every integer $r \ge 5$,*

$$\lceil \log_2 r \rceil + 1 \le si(K_{r,r}) \le \lceil \log_2 r \rceil + 2.$$

Proof. First, suppose that there is a set irregular edge coloring of $K_{r,r}$ using the colors in the set $[k]$ for some integer $k \ge 2$. Since the order of $K_{r,r}$ is $2r$ and there are $2^k - 1$ nonempty subsets of $[k]$, it follows that $2^k - 1 \ge 2r$ and so

$$k \ge \lceil \log_2(2r + 1) \rceil \ge \lceil \log_2(2r) \rceil = \lceil \log_2 r \rceil + 1.$$

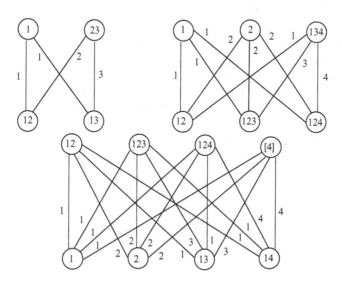

Fig. 4.6 Showing that $si(K_{2,2}) = 3$ and $si(K_{3,3}) = si(K_{4,4}) = 4$

To show that $\text{si}(K_{r,r}) \leq \lceil \log_2 r \rceil + 2$, it suffices to show that there is a set irregular edge coloring of $K_{r,r}$ with the elements in the set $\{1, 2, \ldots, \lceil \log_2 r \rceil + 2\}$. Let $k = \lceil \log_2 r \rceil + 2$ and let t be the unique integer such that $2^{t-1} + 1 \leq r \leq 2^t$, where then $t \geq 3$. Since $t + 1 \leq 2^{t-1}$ for $t \geq 3$, it follows that

$$k = \lceil \log_2 r \rceil + 2 \leq t + 2 \leq 2^{t-1} + 1 \leq r.$$

Let the partite sets of $K_{r,r}$ be $U = \{u_1, u_2, \ldots, u_r\}$ and $W = \{w_1, w_2, \ldots, w_r\}$.

First, we assign each vertex u_i ($1 \leq i \leq r$) a nonempty subset S_i of $[k]$ and assign each vertex w_j ($1 \leq j \leq r$) a nonempty subset T_j of $[k]$ such that $S_1, S_2, \ldots, S_r, T_1, T_2, \ldots, T_r$ are distinct. Let

$$S_1 = \{1, 2\} \quad \text{and} \quad S_i = \{1, 2, i+1\} \quad \text{for } 2 \leq i \leq k-1.$$

Since $k \leq r$, we can choose $S_k, S_{k+1}, \ldots, S_r$ to be any $r - k + 1$ distinct subsets of $[k]$ that contain $\{1, 2\}$ and differ from $S_1, S_2, \ldots, S_{k-1}$. Next, let

$$T_1 = \{1\}, \ T_2 = \{2\}, \ T_j = \{1, j\} \quad \text{for } 3 \leq j \leq k$$

and let $T_{k+1}, T_{k+2}, \ldots, T_r$ be any distinct subsets of $[k]$ that contain 1 but not 2 and differ from $T_1, T_2, \ldots, T_k$. Thus $S_1, S_2, \ldots, S_r, T_1, T_2, \ldots, T_r$ are $2r$ distinct nonempty subsets of $[k]$ and $S_i \cap T_j \neq \emptyset$ for all i, j with $1 \leq i, j \leq r$. We now define an edge coloring $c : E(K_{r,r}) \rightarrow [k]$ by letting $c(u_i w_j)$ to be the maximum element in $S_i \cap T_j$ where $i, j \in \{1, 2, \ldots, r\}$. This is illustrated in Fig. 4.6 for $K_{4,4}$.

It remains to show that $c'(u_i) = S_i$ and $c'(w_j) = T_j$ for $1 \leq i, j \leq r$. By the defining property of c, it follows that $c'(u_i) \subseteq S_i$ and $c'(w_j) \subseteq T_j$ for $1 \leq i, j \leq r$. First, let $p \in S_i$ where $1 \leq i \leq r$. Then $p \in T_p$ and p is the maximum element in $S_i \cap T_p$. Since $u_i w_p \in E(K_{r,r})$, it follows that $c(u_i w_p) = p$ and so $p \in c'(u_i)$. Thus $c'(u_i) = S_i$ for $1 \leq i \leq r$. Next, let $q \in T_j$ where $1 \leq j \leq r$. If $j = 1, 2$, then $S_1 \cap T_j = \{q\}$ and $c(u_1 w_j) = q$, which implies that $q \in c'(w_j)$ and so $c'(w_j) = T_j$. Thus, we may assume that $3 \leq j \leq r$ and so $q \neq 2$. If $q = 1$, then $S_1 \cap T_j = \{1\}$ and $c(u_1 w_j) = 1$; while if $q \geq 3$, then q is the maximum element in $S_{q-1} \cap T_j$ and so $c(u_{p-1} w_j) = q$. In either case, $q \in c'(w_j)$ and so $c'(w_j) = T_j$. Hence, c is vertex-distinguishing, establishing the upper bound. $\qquad \square$

Theorem 4.7 is a special case of the following result.

Theorem 4.8 ([56]). *For positive integers r and s with $r \leq s$,*

$$\text{si}(K_{r,s}) \geq \lceil \log_2(r+s) \rceil.$$

If $r \geq \lceil \log_2 s \rceil + 1$, then

$$\text{si}(K_{r,s}) \leq \lceil \log_2 s \rceil + 2.$$

Chapter 5
Multiset-Defined Irregular Colorings

In Chap. 4, we discussed vertex-distinguishing edge colorings of graphs in which each induced vertex color is the set of colors of its incident edges. We saw that this often requires a large number of colors in comparison with sum-defined vertex colorings described earlier. There is also a related irregular edge coloring for which the induced vertex colors are multisets rather than sets, which is necessarily more restrictive.

5.1 The Multiset Irregular Chromatic Index

Let $c : E(G) \rightarrow [k]$, $k \in \mathbb{N}$, be an unrestricted edge coloring of a graph G that associates with each vertex v of G, the multiset $c'(v)$ of colors of the edges incident with v. Here we commonly write $c'(v)$ as an ordered k-tuple $a_1 a_2 \cdots a_k$, where a_i $(1 \leq i \leq k)$ is the number of edges incident with v that are colored i. Consequently,

$$\sum_{i=1}^{k} a_i = \deg_G v.$$

If $c'(u) \neq c'(v)$ for every two distinct vertices u and v of G, then c is vertex-distinguishing and is called a *multiset irregular edge coloring* of G. If u and v have distinct degrees, then $c'(u) \neq c'(v)$ regardless of how colors are assigned to the edges of G. The minimum number of colors in a multiset irregular edge coloring of G is the *multiset irregular chromatic index* of G, denoted by $\mathrm{mi}(G)$. Thus, $\mathrm{mi}(G) \geq 2$ for every connected graph G of order at least 3. Figure 5.1 shows a graph G with $\mathrm{mi}(G) = 2$ along with a multiset irregular 2-edge coloring of G.

It is immediate that for every connected graph G of order 3 or more,

$$\mathrm{mi}(G) \leq \mathrm{si}(G).$$

© Ping Zhang 2015
P. Zhang, *Color-Induced Graph Colorings*, SpringerBriefs in Mathematics,
DOI 10.1007/978-3-319-20394-2_5

Fig. 5.1 A multiset irregular
2-edge coloring

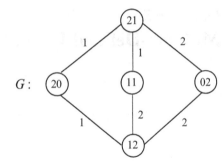

Fig. 5.2 Showing that
$s(P_5) = 3$ and $mi(P_5) = 2$

a

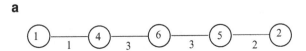

b

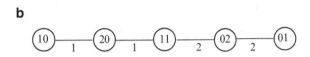

Furthermore, in the sum-defined irregular edge coloring of G introduced in Chap. 2, where the colors are positive integers, the sums of the colors of the incident edges of every two distinct vertices are distinct, which implies that the corresponding multisets are distinct as well. Hence,

$$mi(G) \le s(G)$$

and so distinguishing vertices by means of edge colorings in terms of multisets of edge colors is at least as efficient as distinguishing vertices in terms of sums of edge colors. For example, $s(P_5) = 3$ by Theorem 2.12. (A sum-defined irregular edge coloring of P_5 using the colors 1, 2, 3 is shown in Fig. 5.2a.) On the other hand, the edge coloring shown in Fig. 5.2b using only the colors 1 and 2 is multiset irregular and so $mi(P_5) = 2$.

Multiset irregular edge colorings were introduced and studied by Aigner, Triesch and Tuza in [2, 4] and by Burris in [16, 17]. They referred to these colorings as *irregular colorings* or *vertex-distinguishing edge colorings*. This topic was further studied by Chartrand, Escuadro, Okamoto and Zhang (see [23, 32–36] for example). Multiset irregular edge colorings and the multiset irregular chromatic index of a graph have also been referred to as *detectable labelings* and the *detection number*, respectively, of the graph.

It is sometimes useful to look at multiset irregular edge colorings from another point of view. For a connected graph G of order $n \ge 3$ and a k-tuple factorization $\mathscr{F} = \{F_1, F_2, \cdots, F_k\}$ of G, we associate the ordered k-tuple $a_1 a_2 \cdots a_k$ with a vertex v of G where $\deg_{F_i} v = a_i$ for $1 \leq i \leq k$. Sometimes, such an ordered k-tuple is called a *color code* of the vertex v. Thus $\sum_{i=1}^{k} \deg_{F_i} v = \deg_G v$. If distinct vertices have distinct k-tuples, then we can assign the color i $(1 \leq i \leq k)$ to each edge of F_i and obtain a multiset irregular k-edge coloring c of G for which $c'(v) = a_1 a_2 \cdots a_k$. In this case, the factorization $\mathscr{F}$ is called *irregular*. Conversely, every multiset irregular k-edge coloring of G gives rise to an irregular k-tuple factorization $\mathscr{F} = \{F_1, F_2, \cdots, F_k\}$ of G where the edges of F_i are those edges of G colored i for $1 \leq i \leq k$. Therefore, an edge coloring of a graph G is multiset irregular if and only if the corresponding factorization of G is irregular.

It was stated in Theorem 2.20 that $s(G) \leq n - 1$ for every connected graph G of order $n \geq 4$. Since $\mathrm{mi}(G) \leq s(G)$ for such graphs G, it follows that $\mathrm{mi}(G) \leq n - 1$ as well. A proof that $s(G) \leq n - 1$ was not given because it provided no additional insight into the concept. However, a straightforward proof that $\mathrm{mi}(G) \leq n - 1$ can be given that does not rely on the irregularity strength of G.

Proposition 5.1 ([23]). *If G is a connected graph of order $n \geq 4$, then*

$$\mathrm{mi}(G) \leq n - 1.$$

Proof. We proceed by induction on n. Since the statement is true for the graphs is shown in Fig. 5.3, it follows that the statement holds when $n = 4$.

Assume that $\mathrm{mi}(G) \leq k - 1$ for every connected graph G of a fixed order k where $k \geq 4$ and let H be a connected graph of order $k + 1$. Let v be a vertex of H that is not a cut-vertex. By the induction hypothesis, $\mathrm{mi}(H - v) \leq k - 1$. Hence there exists a multiset irregular $(k - 1)$-edge coloring of $H - v$. Assigning the color k to every edge of H that is incident with v produces a multiset irregular k-edge coloring of H.   □

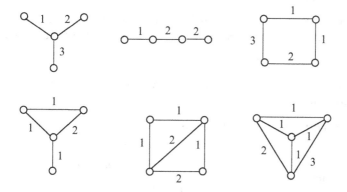

Fig. 5.3 Multiset irregular edge colorings of connected graphs of order 4

5.2 Regular Graphs

It is known that if A is a set containing k different kinds of elements where there are at least r elements of each kind, then the number of different selections of r elements from A is $\binom{r+k-1}{r}$ (see [20, p. 328] for example). In terms of graphs, this says the following.

Theorem 5.2 ([23]). *If there exists a multiset irregular k-edge coloring of a graph G, then at most $\binom{r+k-1}{r}$ vertices of G have degree r.*

In view of Theorem 5.2, if G is an r-regular graph of order n where n is large in comparison to r, then G has a large multiset irregular chromatic index. In particular, for $r \geq 2$, no r-regular graph of order $n \geq 3$ can have multiset irregular chromatic index 2. We verify this fact.

Proposition 5.3 ([17]). *If G is an r-regular graph of order at least 3 where $r \geq 2$, then $\text{mi}(G) \geq 3$.*

Proof. Let c be a 2-edge coloring of G, using the colors 1 and 2, say. If G_1 is the subgraph of G whose edges are colored 1, then G_1 contains two vertices u and v of the same degree, say $\deg_{G_1} u = \deg_{G_1} v = k$. Thus $c'(u) = c'(v) = (k, r - k)$. Therefore, c is not multiset irregular. $\qquad\square$

The following upper bound has been established for the multiset chromatic index of a regular graph.

Theorem 5.4 ([16]). *If G is a connected r-regular graph of order $n \geq 3$, then*

$$\text{mi}(G) \leq (5e(r + 1)!)n)^{\frac{1}{r}},$$

where e is the natural base.

As a consequence of Proposition 5.3 and Theorem 2.4, the complete graphs constitute a class of regular graphs having multiset irregular chromatic index 3.

Theorem 5.5 ([2]). *For every integer $n \geq 3$, $\text{mi}(K_n) = 3$.*

The multiset chromatic indices of cycles are known as well and this information is provided in the next result.

Theorem 5.6 ([23]). *Let $n \geq 3$ be an integer and let $\ell = \left\lceil \sqrt{n/2} \right\rceil$. Then*

$$\text{mi}(C_n) = \begin{cases} 2\ell & \text{if } 2\ell^2 - \ell + 1 \leq n \leq 2\ell^2 \\ 2\ell - 1 & \text{if } 2(\ell - 1)^2 + 1 \leq n \leq 2\ell^2 - \ell. \end{cases}$$

5.3 Complete Bipartite Graphs

By Theorem 2.9, $s(K_{r,r}) = 3$ if $r \geq 2$ is even and $s(K_{r,r}) = 4$ if $r \geq 3$ is odd. Next, we show that $\text{mi}(K_{r,r}) = 3$ for each integer $r \geq 2$.

Theorem 5.7 ([23]). *For every integer $r \geq 2$, $\text{mi}(K_{r,r}) = 3$.*

Proof. By Proposition 5.3 and Theorem 2.9, $\text{mi}(K_{r,r}) = 3$ if r is even. In fact, we show that $K_{r,r}$ has a multiset irregular 3-edge coloring for every integer $r \geq 3$. Let the partite sets of $K_{r,r}$ be $U = \{u_1, u_2, \ldots, u_r\}$ and $W = \{w_1, w_2, \ldots, w_r\}$. Furthermore, let F be the spanning subgraph of $K_{r,r}$ such that $E(F) = \{u_i w_j : i + j \geq r + 1\}$ where $X = \{u_i w_r : 1 \leq i \leq r\}$. Let $\mathscr{F} = \{F_1, F_2, F_3\}$ be the factorization of $K_{r,r}$, where $F_1 = F - X$, F_2 is the complement of F in $K_{r,r}$ and F_3 is the factor of $K_{r,r}$ with $E(F_3) = X$. Observe that

(1) $\deg_{F_2} u_i = \deg_{F_2} w_i = r - i$ for $1 \leq i \leq r$,
(2) $\deg_{F_1} u_r = r - 1$ and $\deg_{F_1} w_r = 0$ and
(3) $\deg_{F_3} u_i = 1$ and $\deg_{F_3} w_i = 0$ for $1 \leq i \leq r - 1$.

Assigning the color i ($i = 1, 2, 3$) to each edge in F_i gives a multiset irregular 3-edge coloring of $K_{r,r}$ and so $\text{mi}(K_{r,r}) = 3$. $\square$

Since the star $K_{1,n-1}$, $n \geq 3$, has $n - 1$ end-vertices, it follows that $\text{mi}(K_{1,n-1}) \geq n - 1$. By Proposition 5.1, $\text{mi}(K_{1,n-1}) = n - 1$. Indeed, the multiset irregular chromatic index of each complete bipartite graph is known.

Theorem 5.8 ([23]). *For integers s and t with $1 \leq s \leq t$ and $s + t \geq 3$,*

$$\text{mi}(K_{s,t}) = \begin{cases} 3 & \text{if } s = t \geq 2 \\ t & \text{if } 1 = s < t \\ 2 & \text{if } t = s + 1 \\ k & \text{if } 2 \leq s < t - 1 \text{ and } k \text{ is the unique integer} \\ & \text{for which } \binom{s+k-2}{s} < t \leq \binom{s+k-1}{s} \end{cases}$$

Proof. Since $\text{mi}(K_{1,t}) = t$ and $\text{mi}(K_{s,s}) = 3$ for $s \geq 2$ by Theorem 5.7, we may assume that $2 \leq s < t$. Let the partite sets of $K_{s,t}$ be

$$U = \{u_1, u_2, \ldots, u_s\} \text{ and } W = \{w_1, w_2, \ldots, w_t\}.$$

First, assume that $t = s + 1$. The coloring that assigns the color 1 to each edge $u_i w_j$ with $i + j \geq s + 2$ and the color 2 to the remaining edges of $K_{s,s+1}$ is multiset irregular and so $\text{mi}(K_{s,s+1}) = 2$. Thus, we now assume that $t \geq s + 2$.

Because $K_{s,t}$ has t vertices of degree s and $t > \binom{s+(k-1)-1}{s}$, it follows from Theorem 5.2 that $\text{mi}(K_{s,t}) \geq k$. To verify that $\text{mi}(K_{s,t}) \leq k$, it suffices to establish the existence of a multiset irregular k-edge coloring of $K_{s,t}$. First, suppose that $t = \binom{s+k-1}{s}$. Then distinct colors $(a_{i1}, a_{i2}, \cdots, a_{ik})$, $1 \leq i \leq t$, can be assigned to the vertices w_i, where $0 \leq a_{ij} \leq k$ for each integer $j \in \{1, 2, \cdots, k\}$ and $\sum_{j=1}^{t} a_{ij} = s$.

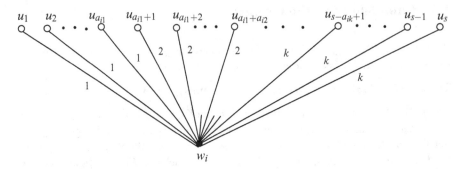

Fig. 5.4 A step in the proof of Theorem 5.8

For $1 \le i \le t$, we color the edges incident with w_i in the order $u_1 w_i, u_2 w_i, \cdots, u_s w_i$. The first a_{i1} of these edges are colored 1, the next a_{i2} of these edges are colored 2 and so on, until we arrive at the last a_{ik} of these edges, which are colored k. In general, for an integer ℓ with $1 \le \ell \le s$, the edge $u_\ell w_i$ is assigned the color $p \in \{1, 2, \cdots, k\}$, where p is the smallest positive integer for which $\ell \le \sum_{j=1}^{p} a_{ij}$. This is illustrated in Fig. 5.4.

If we remove those vertices w_i from $K_{s,t}$ for which $a_{i1} \ne 0$, then the resulting graph is $K_{s,t'}$, where $t' = \binom{s+k-2}{s}$. Let $(b'_{i1}, b'_{i2}, \cdots, b'_{ik})$, $1 \le i \le s$, be the resulting color of u_i in $K_{s,t'}$. From the manner in which the edges of $K_{s,t}$ have been colored, it follows that

$$b'_{1k} < b'_{2k} < \cdots < b'_{sk}.$$

Now suppose that $\binom{s+k-2}{s} < t \le \binom{s+k-1}{s}$. We add back $t - \binom{s+k-2}{s}$ of those vertices w_i for which $a_{i1} \ne 0$ to $K_{s,t'}$ to produce the graph $K_{s,t}$. Let $(b_{i1}, b_{i2}, \cdots, b_{ik})$, $1 \le i \le s$, be the color of u_i in $K_{s,t}$. Since each vertex w_i for which $a_{i1} \ne 0$ is joined to a vertex u_j by an edge colored k only if each edge $w_i u_r$, $j < r \le s$, is also colored k, it follows that

$$b_{1k} < b_{2k} < \cdots < b_{sk}.$$

Hence the vertices of $K_{s,t}$ have distinct color and so this k-coloring of $K_{s,t}$ is multiset irregular. Thus $\mathrm{mi}(K_{s,t}) \le k$, giving the desired result that $\mathrm{mi}(K_{s,t}) = k$. $\qquad\square$

5.4   Trees

In a given tree of order $n \ge 3$, let n_i denote the number of vertices of degree i. For a nontrivial tree, the following identity expresses the number of end-vertices in terms of the number of vertices of higher degrees (see [22, p. 65], for example):

$$n_1 = 2 + n_3 + 2n_4 + 3n_5 + 4n_6 + \cdots . \tag{5.1}$$

The following theorem gives an upper bounds for the irregular chromatic index of a tree in terms of the number of vertices of degree 1 and 2 in the tree.

Theorem 5.9 ([17]). *For a tree T of order at least 3,*

$$\mathrm{mi}(T) \leq \max\{n_1, 4.62\sqrt{n_2}, 8\} + 1,$$

where n_i denotes the number of vertices of T of degree i for $i = 1, 2$.

The exact value of the multiset irregular chromatic index of every path P_n of order $n \geq 3$ is given in the following theorem.

Theorem 5.10 ([33]). *For each integer $n \geq 3$, let $\ell = \left\lceil \dfrac{-3 + \sqrt{8n - 7}}{4} \right\rceil$. Then*

$$\mathrm{mi}(P_n) = \begin{cases} 2\ell & \text{if } 2\ell^2 - \ell + 2 \leq n \leq 2\ell^2 + 3 \\ 2\ell + 1 & \text{if } 2\ell^2 + 4 \leq n \leq 2\ell^2 + 3\ell + 2. \end{cases}$$

It's a consequence of Theorem 5.2 that if T is a tree with multiset irregular chromatic index k, then T contains at most k end-vertices and at most $\frac{k(k+1)}{2}$ vertices of degree 2, that is, $n_1 \leq k$ and $n_2 \leq \frac{k(k+1)}{2}$. Solving $n_2 = \frac{k(k+1)}{2}$ for k, we obtain

$$k = \frac{-1 + \sqrt{8n_2 + 1}}{2}.$$

Thus if T is a tree of order at least 3 with n_1 end-vertices and n_2 vertices of degree 2, then

$$\mathrm{mi}(T) \geq \max\left\{n_1, \left\lceil \frac{-1 + \sqrt{8n_2 + 1}}{2} \right\rceil\right\}.$$

In [35], the problem is investigated of how large and how small the multiset irregular chromatic index of a tree of order n can be. For each integer $n \geq 3$, let $D_T(n)$ denote the maximum multiset irregular chromatic index among all trees of order n and $d_T(n)$ the minimum multiset irregular chromatic index among all trees of order n. That is, if $\mathscr{T}_n$ is the set of all trees of order n, then

$$D_T(n) = \max\{\mathrm{mi}(T) : T \in \mathscr{T}_n\}$$
$$d_T(n) = \min\{\mathrm{mi}(T) : T \in \mathscr{T}_n\}.$$

These concepts were introduced by Escuadro and Zhang in 2008.

Theorem 5.11 ([35]). *For each integer $n \geq 3$,*

$$D_T(n) = n - 1 \quad \text{and} \quad d_T(n) = \left\lceil \frac{-5 + \sqrt{8n + 41}}{2} \right\rceil.$$

Furthermore, for each pair k, n of integers with $n \geq 3$ and $d_T(n) \leq k \leq D_T(n)$, there exists a tree T of order n with $\mathrm{mi}(T) = k$.

5.5   Max-Min Value Problems

If G is a connected graph of order n and size m, then the number of edges that must be deleted from G to obtain a spanning tree of G is $m - n + 1$. The number $m - n + 1$ is referred to as the *cycle rank* of G. Consequently, if G is a connected graph of order n, size m and cycle rank ψ, then

$$m = (n - 1) + \psi \leq \binom{n}{2},$$

which implies that

$$n \geq \left\lceil \frac{3 + \sqrt{1 + 8\psi}}{2} \right\rceil.$$

For integers ψ and n, where $\psi \geq 0$ and $n \geq \left\lceil \frac{3 + \sqrt{1+8\psi}}{2} \right\rceil$, let $D_\psi(n)$ denote the maximum multiset irregular chromatic index among all connected graphs of order n with cycle rank ψ and let $d_\psi(n)$ denote the minimum multiset irregular chromatic index among all connected graphs of order n with cycle rank ψ. Hence, if $\mathscr{G}_{\psi,n}$ denotes the set of all connected graphs of order n with cycle rank ψ, then

$$D_\psi(n) = \max \{mi(G) : G \in \mathscr{G}_{\psi,n}\}$$
$$d_\psi(n) = \min \{mi(G) : G \in \mathscr{G}_{\psi,n}\}.$$

It follows that $D_0(n) = D_T(n)$ is the maximum multiset irregular chromatic index among all trees of order n, $D_1(n)$ is the maximum multiset irregular chromatic index among all unicyclic graphs (connected graphs with exactly one cycle) of order n and $D_2(n)$ is the maximum multiset irregular chromatic index among all connected graphs of order n whose cycle rank is 2. Furthermore, $d_0(n) = d_T(n), d_1(n)$ and $d_2(n)$ are the minimum multiset irregular chromatic indices among all trees, unicyclic graphs and connected graphs with cycle rank 2 of order n, respectively. Formulas have been established for $D_\psi(n)$ and $d_\psi(n)$ when $\psi = 1, 2$.

Theorem 5.12 ([33]). *For each integer $n \geq 3$,*

- $D_1(n) = 3$ *for* $n = 3, 4, 5$ *and* $D_1(n) = n - 3$ *for* $n \geq 6$,

- $d_1(3) = 3$ *and* $d_1(n) = \left\lceil \dfrac{-5 + \sqrt{8n + 41}}{2} \right\rceil$ *for* $n \geq 4$.

Furthermore, for each pair k, n of integers with $n \geq 3$ and $d_1(n) \leq k \leq D_1(n)$, there exists a unicyclic graph G of order n with $\det(G) = k$.

Theorem 5.13 ([33]). *For each integer $n \geq 4$,*

- $D_2(4) = 2$, $D_2(n) = 3$ *for* $n = 5, 6, 7$ *and* $D_2(n) = n - 4$ *for* $n \geq 8$,

- $d_2(n) = 2$ *for* $4 \leq n \leq 9$ *and* $d_2(n) = \left\lceil \dfrac{-5 + \sqrt{8n + 9}}{2} \right\rceil$ *for* $n \geq 10$.

Furthermore, for each pair k, n of integers with $n \geq 4$ and $d_2(n) \leq k \leq D_2(n)$, there exists a connected graph G of order n having cycle rank 2 and $\det(G) = k$.

For integers ψ, t, n with $t \geq 3$, $n \geq t + 3$ and $\binom{t-2}{2} + 1 \leq \psi \leq \binom{t-1}{2}$, it is possible to construct a connected graph G of order n having cycle rank ψ such that $\mathrm{mi}(G) \geq n - t$. For example, let T be a tree of order n with exactly $n - t$ end-vertices and let $V(T) = U_1 \cup U_2$, where U_1 is the set of end-vertices of T and $U_2 = V(T) - U_1$. Thus $|U_2| = t$, the subgraph $T[U_2]$ induced by U_2 is connected and $|E(T[U_2])| = t - 1$. Since

$$\psi \leq \binom{t-1}{2} = \binom{t}{2} - (t - 1),$$

we can construct a connected graph $G \in \mathscr{G}_{\psi, n}$ from T by adding ψ edges to the vertices of U_2 in T. Since G contains exactly $n - t$ end-vertices, $\mathrm{mi}(G) \geq n - t$. These observations yield the following.

Proposition 5.14 ([32]). *If $\psi \geq 1$, $t \geq 3$ and $n \geq t + 3$ are integers with*

$$\binom{t-2}{2} + 1 \leq \psi \leq \binom{t-1}{2},$$

then $D_\psi(n) \geq n - t$.

Let G be a nontrivial connected graph having maximum degree Δ and cycle rank ψ. Recall that if n_i is the number of vertices of degree i in G where $1 \leq i \leq \Delta$, then

$$n_1 = (2 - 2\psi) + n_3 + 2n_4 + 3n_5 + \cdots + (\Delta - 2)n_\Delta$$

Furthermore, if $\mathrm{mi}(G) = k$, then G has at most k end-vertices and at most $\frac{k^2 + k}{2}$ vertices of degree 2. Since

$$n_1 = (2 - 2\psi) + n_3 + 2n_4 + 3n_5 + \cdots + (\Delta - 2)n_\Delta$$
$$\geq (2 - 2\psi) + n_3 + n_4 + n_5 + \cdots + n_\Delta,$$

it follows that

$$n_3 + n_4 + n_5 + \cdots + n_\Delta \leq n_1 - (2 - 2\psi) \leq k + 2\psi - 2$$

and so

$$n \leq k + \frac{k^2 + k}{2} + (k + 2\psi - 2) = \frac{k^2 + 5k + 4\psi - 4}{2}.$$

Hence the largest possible order of a connected graph having cycle rank ψ and multiset irregular chromatic index k is $\frac{k^2 + 5k + 4\psi - 4}{2}$. That is, if G is a connected graph of order n with

$$\frac{(k-1)^2 + 5(k-1) + 4\psi - 4}{2} + 1 \leq n \leq \frac{k^2 + 5k + 4\psi - 4}{2}$$

and having cycle rank ψ, then $\mathrm{mi}(G) \geq k$, implying that $d_\psi(n) \geq k$. As a consequence of these observations, we have the following.

Proposition 5.15 ([32]). *For integers $\psi \geq 1$ and $n \geq 2 + 2\psi$,*

$$d_\psi(n) \geq \left\lceil \frac{-5 + \sqrt{8n + (41 - 16\psi)}}{2} \right\rceil.$$

Note that for Proposition 5.15 to be nontrivial, we assume that

$$\left\lceil \frac{-5 + \sqrt{8n + (41 - 16\psi)}}{2} \right\rceil \geq 2$$

and so $n \geq 2 + 2\psi$ as stated in the result.

We conclude this section with the following conjectures.

Conjecture 5.16 ([32]). *If $\psi \geq 1$, $t \geq 3$ and $n \geq t + 3$ are integers with*

$$\binom{t-2}{2} + 1 \leq \psi \leq \binom{t-1}{2},$$

then $D_\psi(n) = n - t$.

Conjecture 5.17 ([32]). *For integers $\psi \geq 1$ and $n \geq 2 + 2\psi$,*

$$d_\psi(n) = \left\lceil \frac{-5 + \sqrt{8n + (41 - 16\psi)}}{2} \right\rceil.$$

Chapter 6
Sum-Defined Neighbor-Distinguishing Colorings

We now switch our attention from edge colorings of a graph G that give rise to vertex-distinguishing vertex colorings to those that result in neighbor-distinguishing vertex colorings. That is, suppose that $c : E(G) \to [k]$ for some positive integer k is an unrestricted edge coloring. For each vertex v of G, we assign a color $c'(v)$ to v that depends in some way on the colors of the edges incident with v. The coloring c' is *neighbor-distinguishing* if $c'(u) \neq c'(v)$ for every two adjacent vertices u and v of G. That is, c' is a proper vertex coloring. If the edge coloring c induces a neighbor-distinguishing vertex coloring of G, then c is referred to as a *neighbor-distinguishing edge coloring* of G. A number of neighbor-distinguishing edge colorings have been introduced and studied. We look at one of these in this chapter.

6.1 The Sum Distinguishing Index

In 2004 a neighbor-distinguishing edge coloring $c : E(G) \to \{1, 2, \ldots, k\}$ of a connected graph G of order at least 3 was introduced (see [19, p. 385]), where $k \in \mathbb{N}$, in which an induced vertex coloring $c' : V(G) \to \mathbb{N}$ is defined by

$$c'(v) = \sum_{e \in E_v} c(e)$$

for each $v \in V(G)$. If $c'(x) \neq c'(y)$ for every pair x, y of adjacent vertices of G, then c is called a *proper sum k-edge coloring*. The minimum k for which a graph G has a proper sum k-edge coloring is the *proper sum neighbor-distinguishing chromatic index* (or, more simply, *sum distinguishing index*) and is denoted by $sd(G)$ of G. A proper sum $sd(G)$-edge coloring of G is a *minimum proper sum coloring* of G.

© Ping Zhang 2015
P. Zhang, *Color-Induced Graph Colorings*, SpringerBriefs in Mathematics,
DOI 10.1007/978-3-319-20394-2_6

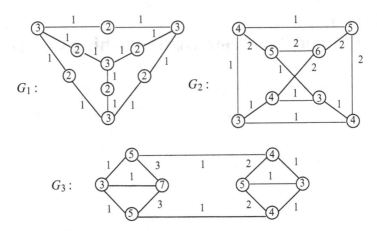

Fig. 6.1 Minimum proper sum colorings of graphs

Since the sum distinguishing index of a connected graph of order at least 3 never exceeds the irregularity strength of the graph, it follows that $sd(G) \le s(G)$ for each such graph G.

Figure 6.1 shows three graphs G_1, G_2 and G_3, where $sd(G_i) = i$ for $i = 1, 2, 3$. A minimum proper sum coloring is given in each case. Furthermore, the graph G_1 in Fig. 6.1 illustrates the following observation.

Observation 6.1. *A connected graph G of order at least 3 has sum distinguishing index 1 if and only if the degrees every two adjacent vertices of G are different.*

To illustrate these concepts, we determine the sum distinguishing indices of complete graphs and complete bipartite graphs. In Theorem 2.4, it was shown that $s(K_n) = 3$ for every integer $n \ge 3$. Since every two vertices in K_n are adjacent, it follows that every neighbor-distinguishing edge coloring of K_n is also a vertex-distinguishing edge coloring of K_n. Therefore, we have the following.

Proposition 6.2. *For each integer $n \ge 3$, $sd(K_n) = 3$.*

The sum distinguishing indices of complete bipartite graphs are given below.

Proposition 6.3. *For positive integers s and t with $s + t \ge 3$,*

$$sd(K_{s,t}) = \begin{cases} 1 & \text{if } s \ne t \\ 2 & \text{if } s = t. \end{cases}$$

Proof. If $s \ne t$, then no two adjacent vertices of $K_{s,t}$ have the same degree and so $sd(K_{s,t}) = 1$. Suppose then that $s = t$. Therefore, $sd(K_{s,s}) \ge 2$. Let the partite sets of $K_{s,s}$ be $U = \{u_1, u_2, \cdots, u_s\}$ and $W = \{w_1, w_2, \cdots, w_s\}$, and let c be the 2-edge coloring of $K_{s,s}$ defined by

$$c(e) = \begin{cases} 1 & \text{if } e = u_1 w_i \text{ for } 1 \le i \le s \\ 2 & \text{otherwise.} \end{cases}$$

Then

$$c'(v) = \begin{cases} 2 & \text{if } v = u_1 \\ 2s & \text{if } v = u_i \text{ for } 2 \le i \le s \\ 2s - 1 & \text{if } v = w_i \text{ for } 1 \le i \le s. \end{cases}$$

For every two adjacent vertices of $K_{s,s}$, one has an even color and the other has an odd color. It follows therefore that c is a neighbor-distinguishing 2-edge coloring of $K_{s,s}$ and so $sd(K_{s,s}) = 2$. ☐

6.2 The 1-2-3 Conjecture

Of all the graphs G we have considered thus far, either $sd(G) = 1$, $sd(G) = 2$ or $sd(G) = 3$. In fact, Karoński et al. [67] conjectured that these are the only three possibilities for every connected graph of order at least 3. This conjecture goes by a rather catchy name.

The 1-2-3 Conjecture. If G is a connected graph of order 3 or more, then $sd(G)$ has one of the values 1, 2, 3.

Consequently, if the 1-2-3 Conjecture is true, then for every connected graph G of order 3 or more, it is possible to assign each edge of G one of the colors 1, 2, 3 in such a way that for every two adjacent vertices of G, the sums of the colors of their incident edges are different. This conjecture was verified for all 3-colorable graphs by Karoński et al. [67]. In the proof of this fact that follows, it is useful to observe that since every connected 3-chromatic graph G contains an odd cycle, there is a path of even length connecting every two vertices of G.

Theorem 6.4. *Let G be a connected graph of order 3 or more. If $\chi(G) \le 3$, then $sd(G)$ has one of the values 1, 2, 3.*

Proof. First, we introduce some notation and terminology. For every 3-edge coloring of G using colors from the set $\{1, 2, 3\}$ and for each vertex v of G, let $\sigma(v)$ denote the sum of the colors modulo 3 of the edges incident with v such that $\sigma(v) \in \{1, 2, 3\}$. We refer to $\sigma(v)$ as the *color sum* of v. We now consider two cases, according to whether $\chi(G) = 3$ or $\chi(G) = 2$, that is, G is bipartite.

Case 1. $\chi(G) = 3$. Since G is 3-chromatic, there exists a proper vertex 3-coloring f of G using the colors 1, 2 and 3. Suppose that n_i vertices of G are colored i for $i = 1, 2, 3$. Thus the order of G is $n = n_1 + n_2 + n_3$. Then $n_1 + 2n_2 + 3n_3$ equals 1, 2 or 3 modulo 3 or equivalently, 4, 2 or 6 modulo 3. That is, $n_1 + 2n_2 + 3n_3 \equiv 2j$ (mod 3) for some $j \in \{1, 2, 3\}$. Let c be an edge coloring of G in which some edge of G is assigned the color j and all other edges of G are assigned the color 3. Then the sum of the color sums of the vertices of G is congruent to $2j$ modulo 3, that is,

$$\sum_{v \in V(G)} \sigma(v) \equiv 2j \pmod{3}.$$

We now modify the edge coloring of G so that $\sum_{v \in V(G)} \sigma(v)$ is unchanged and such that $\sigma(v) = f(v)$ for each $v \in V(G)$. Suppose that there is a vertex u of G such that $\sigma(u) \neq f(u)$. Since $\sum_{v \in V(G)} \sigma(v) \equiv 2j \pmod{3}$ and $n_1 + 2n_2 + 3n_3 = \sum_{v \in V(G)} f(u) \equiv 2j \pmod{3}$, it follows that $\sum_{v \in V(G)} (\sigma(v) - f(v)) \equiv 0 \pmod{3}$ and so there is another vertex w such that $\sigma(w) \neq c(w)$. Let W be a $u - w$ walk of even length in G. Adding the colors $f(u) - \sigma(u)$, $\sigma(u) - f(u)$, $f(u) - \sigma(u)$, ..., $\sigma(u) - f(u)$ alternately to the edges of W gives us a new value of $\sigma(u)$, namely $\sigma(u) = f(u)$ and does not change the value of $\sigma(x)$ for any other vertex x except w, and leaves $\sum_{v \in V(G)} \sigma(v)$ unchanged. Repeated application of this gives the desired edge coloring, that is, $\sigma(v) = f(v)$ for each $v \in V(G)$.

Case 2. G is bipartite. Let there be given a proper vertex 2-coloring f of G using the colors 1 and 2 modulo 3 such that the color 1 is assigned to a vertex x of degree 2 or more. Now let there be given an edge coloring of G that assigns each edge of G the color 3 modulo 3. Then $\sigma(v) = 3$ for all vertices v of G. We now modify the edge coloring of G so that $\sigma(t) \neq 3$ if $f(t) = 1$ and $\sigma(t) = 3$ if $f(t) = 2$, which will result in every two adjacent vertices of G having different color sums.

Suppose first that there are three or more vertices of G colored 1. Let u, v and w be three such vertices. We then add the colors $1, 2, 1, \ldots, 2$ alternately to the edges of a $u - v$ path in G. Then $\sigma(u) = 1$ and $\sigma(v) = 2$, where $\sigma(t)$ is unchanged for all other vertices t in G. Next, we add the colors $2, 1, 2, \ldots, 1$ alternately to the edges of a $v - w$ path in G. Then $\sigma(u) = \sigma(v) = \sigma(w) = 1$ and $\sigma(t)$ is unchanged for all other vertices t in G. We continue this procedure as long as there are three or more vertices colored 1 whose color sums are 3. If no such vertices remain, then the resulting 3-edge coloring of G has the desired property. Hence, we may assume that either exactly two such vertices remain or exactly one such vertex remains. We consider these two subcases.

Subcase 2.1. G contains exactly two vertices u and v colored 1 with $\sigma(u) = \sigma(v) = 3$, where $\sigma(t) = 1$ if $f(t) = 1$ and $\sigma(t) = 3$ if $f(t) = 2$ for all $t \in V(G) - \{u, v\}$. We add the colors $1, 2, 1, \ldots, 2$ alternately to the edges of a $u - v$ path in G, resulting in $\sigma(u) = 1$ and $\sigma(v) = 2$, where $\sigma(t)$ is unchanged for all other vertices t in G. Thus, the resulting 3-edge coloring of G has the desired property.

Subcase 2.2. G contains exactly one vertex y colored 1 with $\sigma(y) = 3$, where again $\sigma(t) = 1$ if $f(t) = 1$ and $\sigma(t) = 3$ if $f(t) = 2$ for all $t \in V(G) - \{y\}$. We may assume that the vertex y is x, for if not, we may alternately add the colors $2, 1, 2, \ldots, 1$ to the edges of an $x - y$ path, resulting in $\sigma(x) = 3$ and $\sigma(y) = 1$. We now add the color 2 to any two edges incident with x, say xp and xq, obtaining $\sigma(x) = 1$ and $\sigma(p) = \sigma(q) = 2$, where $\sigma(t)$ is unchanged for all other vertices t in G. Furthermore, p and q are not adjacent since G is bipartite and $\sigma(z) = 1$ for each vertex z different from x that is adjacent to p or q. Hence, $\sigma(u) \neq \sigma(v)$ for every two adjacent vertices u and v of G.

Therefore, there is a 3-edge coloring of G such that every two adjacent vertices of G have different color sums and so different color codes, which implies that $sd(G) \leq 3$.   □

6.3   The Multiset Distinguishing Index

It was mentioned in Chap. 5 that for every connected graph G of order at least 3, the multiset irregular chromatic index of G is at most the irregularity strength of G—in symbols, $mi(G) \leq si(G)$. Furthermore, it was stated that this inequality can be strict since, for example, $mi(P_5) = 2$ and $s(P_5) = 3$.

There is a neighbor-distinguishing analogue of the multiset irregular chromatic index. An edge coloring c of a connected graph G of order at least 3 is called *multiset neighbor-distinguishing* if the multisets of colors of the incident edges of every two adjacent vertices are different. The minimum number of colors in such an edge coloring is called the *multiset neighbor-distinguishing chromatic index* of G (also called the *multiset distinguishing index*), denoted by $md(G)$. Just as in the case for the corresponding parameters $mi(G)$ and $s(G)$, we have

$$md(G) \leq sd(G)$$

for every connected graph G of order at least 3.

Since $md(G) = sd(G) = 1$ if and only if every two adjacent vertices of G have different degrees, it follows that if $sd(G) = 2$, then $md(G) = 2$ as well. The 2-edge coloring of the tree T in Fig. 6.2 is multiset neighbor-distinguishing but it is not a proper sum edge coloring since the sums of the colors of the incident edges of u and v are both 5. Nevertheless, $md(T) = sd(T) = 2$.

Indeed, we know of no graphs G for which $md(G) < sd(G)$, which brings up the following problem.

Question 6.5.   Does there exist a connected graph G of order at least 3 such that

$$md(G) < sd(G)?$$

If the 1-2-3 Conjecture is true and the answer to the question stated in Question 6.5 is yes, then a graph G with $md(G) < sd(G)$ likely has the values $md(G) = 2$ and $sd(G) = 3$.

Fig. 6.2   A multiset neighbor-distinguishing of a tree that is not a proper sum edge coloring

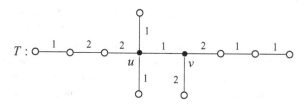

By Theorem 6.4, $sd(G) \leq 3$ for every graph G with $\chi(G) \leq 3$. Therefore, $md(G) \leq 3$ as well if $\chi(G) \leq 3$. If $\chi(G) \geq 4$, then the best that has been shown is that $md(G) \leq 4$. To establish this result, we first verify a lemma due to Addario-Berry et al. [1].

Lemma 6.6. *If G is a connected graph with $\chi(G) \geq 4$, then there exists a partition of $V(G)$ into three sets V_1, V_2 and V_3 such that for $i = 1, 2, 3$,*

(1) $|N(v) \cap V_{i+1}| \geq |N(v) \cap V_i|$ *for each $v \in V_i$ and*
(2) *every vertex in V_i has a neighbor in V_{i+1} (where $V_4 = V_1$).*

Proof. Among all partitions of $V(G)$ into three sets, let $\mathscr{P} = \{U_1, U_2, U_3\}$ be one for which the number of edges joining vertices in different subsets in $\mathscr{P}$ is maximum. We refer to such a partition of $V(G)$ as a *maximum 3-partition*. We claim that condition (1) holds. Suppose that it does not. Then we may assume that G contains a vertex u in U_1, say, such that $|N(u) \cap U_2| < |N(u) \cap U_1|$. Let $P' = \{U_1', U_2', U_3'\}$ be the partition of $V(G)$ where $U_1' = U_1 - \{u\}$, $U_2' = U_2 \cup \{u\}$ and $U_3' = U_3$. Since the number of edges joining vertices in different subsets in $\mathscr{P}'$ exceeds that in $\mathscr{P}$, the assumption that $\mathscr{P}$ is a maximum 3-partition is contradicted. Thus, as claimed, every maximum 3-partition of $V(G)$ satisfies (1).

We now show that there exists a maximum 3-partition of $V(G)$ that satisfies condition (2). For each maximum 3-partition $\mathscr{P} = \{U_1, U_2, U_3\}$ of $V(G)$, define a digraph $D_{\mathscr{P}}$ such that $V(D_{\mathscr{P}}) = V(G)$ and $E(D_{\mathscr{P}})$ contains (i) the arc (u, w) for each edge $uw \in E(G)$ with $u \in U_i$ and $w \in U_{i+1}$, where $i \in \{1, 2, 3\}$ and (ii) the symmetric arcs (u, w) and (w, u) for each edge $uw \in E(G)$, where $u, w \in U_i$ for some $i \in \{1, 2, 3\}$.

Consequently, in $D_{\mathscr{P}}$ a vertex $u \in U_i$ has od $u \geq 1$ in $D_{\mathscr{P}}$ if and only if u has a neighbor in $U_i \cup U_{i+1}$. Since $\mathscr{P}$ satisfies condition (1), we see that if u has a neighbor in U_i, then u has a neighbor in U_{i+1} as well. Hence if all vertices in $D_{\mathscr{P}}$ have outdegree 1 or more, then condition (2) is also satisfied.

A vertex x is said to be a *descendant* of a vertex u if $D_{\mathscr{P}}$ contains a directed $u - x$ path. Let $S_{\mathscr{P}}$ be the set of all vertices u for which either

(a) u belongs to a directed cycle in $D_{\mathscr{P}}$ (including a directed 2-cycle) or
(b) u has a descendant x that belongs to a directed cycle in $D_{\mathscr{P}}$.

Consequently, every vertex in $S_{\mathscr{P}}$ has outdegree at least 1 in $D_{\mathscr{P}}$. Therefore, if $S_{\mathscr{P}} = V(G)$, then condition (2) holds. Suppose then that $S_{\mathscr{P}} \neq V(G)$ for every maximum 3-partition $\mathscr{P}$ of $V(G)$. Among all maximum 3-partitions of $V(G)$, assume that $\mathscr{P}$ is one for which $|S_{\mathscr{P}}|$ is maximum. We then construct a new partition $\mathscr{P}' = \{U_1', U_2', U_3'\}$ of $V(G)$ from $\mathscr{P}$ by transferring each vertex not in $S_{\mathscr{P}}$ to the succeeding subset in $\mathscr{P}$, that is, if $y \in S_{\mathscr{P}} \cap U_i$, then $y \in U_i'$; while if $y \in (V(G) - S_{\mathscr{P}}) \cap U_i$, then $y \in U_{i+1}'$. We claim that $\mathscr{P}'$ is also a maximum 3-partition of $V(G)$. To verify this, we show that any edge joining vertices in different subsets of $\mathscr{P}$ joins vertices in different subsets in $\mathscr{P}'$.

Any two vertices belonging to $S_{\mathscr{P}}$ that join vertices in two different sets in $\mathscr{P}$ have the same property in $\mathscr{P}'$. Hence it suffices to consider adjacent vertices y and

z at least one of which does not belong to $S_{\mathscr{P}}$. If $y \in U_i - S_{\mathscr{P}}$ where $i \in \{1, 2, 3\}$, then $z \notin U_i$, for otherwise both (y, z) and (z, y) are arcs in $D_{\mathscr{P}}$, which would contradict (a). Thus either (y, z) or (z, y) is an arc of $D_{\mathscr{P}}$ but not both, say (y, z) is an arc of $D_{\mathscr{P}}$. Thus, if $y \in U_i - S_{\mathscr{P}}$, then $z \in U_{i+1}$ where $i \in \{1, 2, 3\}$. Furthermore, if $z \in S_{\mathscr{P}}$, then $y \in S_{\mathscr{P}}$ as well (since (y, z) is an arc of $D_{\mathscr{P}}$). Therefore, there are only two cases remaining to consider.

Case 1. $y, z \notin S_{\mathscr{P}}$. Then $y \in U_i$ and $z \in U_{i+1}$ for some $i \in \{1, 2, 3\}$, which implies that $y \in U'_{i+1}$ and $z \in U'_{i+2}$.

Case 2. $y \in S_{\mathscr{P}}$ and $z \notin S_{\mathscr{P}}$. Then $y \in U_i$ and $z \in U_{i+1}$ for some $i \in \{1, 2, 3\}$. This implies that $y \in U'_i$ and $z \in U'_{i+2}$. In this case, the arc (z, y) is in $D_{\mathscr{P}'}$.

Thus, as claimed, $\mathscr{P}'$ is a maximum 3-partition of $V(G)$.

Since $\chi(G) \geq 4$, not all of the sets U_1, U_2 and U_3 in $\mathscr{P}$ can be independent. Hence at least one of these sets contains two adjacent vertices, implying that $D_{\mathscr{P}}$ contains a directed 2-cycle and that $S_{\mathscr{P}} \neq \emptyset$. Since G is connected and $S_{\mathscr{P}} \neq V(G)$, it follows that G contains an edge uw such that $u \in S_{\mathscr{P}}$ and $w \notin S_{\mathscr{P}}$. In particular, this says that u cannot be a descendant of w. Thus (w, u) cannot be an arc in $D_{\mathscr{P}}$ and so (u, w) is an arc in $D_{\mathscr{P}}$.

We now show that (i) every vertex z belonging to $S_{\mathscr{P}}$ also belongs to $S_{\mathscr{P}'}$ and (ii) $w \in S_{\mathscr{P}'}$. To verify (i), let $z \in S_{\mathscr{P}}$. Hence either z lies on a directed cycle in $D_{\mathscr{P}}$ or there is a directed $z - x$ path in $D_{\mathscr{P}}$ where x is on a directed cycle in $D_{\mathscr{P}}$. Thus all vertices on this cycle or path belong to $S_{\mathscr{P}}$ as well. Since the only arcs that are reversed in $D_{\mathscr{P}'}$ have one of its incident vertices not in $S_{\mathscr{P}}$, it follows that the cycle or path in $D_{\mathscr{P}}$ is unchanged in $D_{\mathscr{P}'}$, implying that $z \in S_{\mathscr{P}'}$. We now verify (ii). As we saw in Case 2, the arc in $D_{\mathscr{P}'}$ corresponding to the arc (u, w) in $D_{\mathscr{P}}$ is (w, u). Since $u \in S_{\mathscr{P}'}$, it follows that $w \in S_{\mathscr{P}'}$. Hence $|S_{\mathscr{P}'}| \geq |S_{\mathscr{P}}| + 1$, which contradicts our assumption that $|S_{\mathscr{P}}|$ is maximized over all maximum 3-partitions $\mathscr{P}$ of $V(G)$. □

With the aid of Lemma 6.6, Addario-Berry et al. [1] showed that every connected graph of order 3 or more has multiset distinguishing index at most 4.

Theorem 6.7. *If G is a connected graph of order 3 or more, then $md(G)$ has one of the values 1, 2, 3, 4.*

Proof. By Theorem 6.4, $sd(G) \leq 3$ if $\chi(G) \leq 3$. Hence we may assume that $\chi(G) \geq 4$. By Lemma 6.6, there exists a partition $\mathscr{P} = \{V_1, V_2, V_3\}$ of $V(G)$ satisfying the following two conditions:

(1) $|N(v) \cap V_{i+1}| \geq |N(v) \cap V_i|$ for each $v \in V_i$ and
(2) every vertex in V_i has a neighbor in V_{i+1} (where $V_4 = V_1$).

We now assign one of the colors 1, 2, 3, 4 to each edge of G in the following manner. For $i \in \{1, 2, 3\}$, each edge joining two vertices of V_i is colored i. Each edge joining a vertex in V_i and a vertex in V_{i+1} will be colored either i or 4 with the condition that for each vertex $v \in V_i$ for which $N(v) \cap V_i = \emptyset$, each edge of

G joining v and a vertex of V_{i+1} is assigned the color i. (By (2), it follows that if $N(v) \cap V_i = \emptyset$, there is at least one edge joining v and a vertex of V_{i+1}.)

For $i \in \{1, 2, 3\}$, we now consider all those vertices $v \in V_i$ for which $|N(v) \cap V_i| \geq 1$. For each such vertex v, let $|N(v) \cap V_i| = k_v$ and so $k_v \geq 1$. Let $v_1, v_2, \ldots, v_s$ be an ordering of all such vertices in V_i. We now assign to each vertex v_j ($1 \leq j \leq s$) a color $f(v_j)$ that will be used to determine the number of edges joining v_j and the vertices in V_{i+1} that will be colored i. Color the vertex v_1 with $f(v_1) = k_{v_1}$. For $2 \leq r \leq s$, suppose that the vertices $v_1, v_2, \ldots, v_{r-1}$ have been assigned colors, namely $f(v_1), f(v_2), \ldots, f(v_{r-1})$. We then define the color $f(v_r)$ as the smallest integer at least k_{v_r} that is distinct from the colors that have already been assigned to the neighbors of v_r in V_i. For each vertex $v_j \in V_i$, where $1 \leq j \leq s$, we now assign the color i to $f(v_j) - k_{v_j}$ edges joining v_j and the vertices in V_{i+1} and the color 4 to the remaining edges.

Now consider two adjacent vertices u and v belonging to different sets in $\mathscr{P}$, say $u \in V_i$ and $v \in V_{i+1}$. Since v is incident with an edge colored $i+1$ while u is not, it follows that u and v have different color codes. Hence it suffices to show that every two adjacent vertices belonging to the same set V_i have distinct color codes.

For each j with $1 \leq j \leq s$, the vertex v_j is incident with exactly k_{v_j} edges within V_i that are colored i and is incident with exactly $f(v_j) - k_{v_j}$ edges between V_i and $V(G) - V_i$ that are colored i. Thus each vertex v_j ($1 \leq j \leq s$) is incident with exactly $f(v_j)$ edges colored i. From the manner in which the colors $f(v_j)$ are defined, any two vertices in $\{v_1, v_2, \ldots, v_s\}$ have different colors and so have different color codes. $\qquad\square$

Despite Theorem 6.7, no graph G is known for which $sd(G) > 3$. There are many papers dealing with this topic (see [37, 65, 74], for example).

Chapter 7
Modular Sum-Defined Neighbor-Distinguishing Colorings

While Chap. 2 dealt with sum-defined vertex-distinguishing edge colorings whose colors are positive integers, Chap. 3 considered corresponding edge colorings whose colors are elements of the integers modulo k for some integer $k \geq 2$. On the other hand, Chap. 6 concerned sum-defined neighbor-distinguishing edge colorings whose colors are positive integers. In the current chapter, we turn our attention to sum-defined neighbor-distinguishing edge colorings where the colors are elements of $\mathbb{Z}_k$ for integers $k \geq 2$.

7.1 Modular Chromatic Index

For a connected graph G of order n at least 3, let $c : E(G) \rightarrow \mathbb{Z}_k$ $(k \geq 2)$ be an unrestricted edge coloring of G. Here, the color $c'(v)$ of a vertex v of G is defined as the sum in $\mathbb{Z}_k$ of the colors of the edges incident with v; that is, for the set E_v of edges incident with v in G,

$$c'(v) = \sum_{e \in E_v} c(e).$$

The edge coloring c is a *modular neighbor-distinguishing k-edge coloring* of G if $c'(u) \neq c'(v)$ for all pairs u, v of adjacent vertices of G. We refer to such edge colorings more simply as *modular k-edge colorings*. An edge coloring c is a *modular edge coloring* if c is a modular k-edge coloring for some integer $k \geq 2$. The *modular chromatic index* $\chi'_m(G)$ of G is the minimum k for which G has a modular k-edge coloring. Recall that the modular edge-gracefulness $\operatorname{meg}(G)$ of a graph G is the minimum number of colors required of a vertex-distinguishing modular unrestricted edge coloring of G. Since the modular edge-gracefulness $\operatorname{meg}(G)$ of G exists for every connected graph G of order at least 3, it follows that $\chi'_m(G)$ exists as well.

© Ping Zhang 2015
P. Zhang, *Color-Induced Graph Colorings*, SpringerBriefs in Mathematics,
DOI 10.1007/978-3-319-20394-2_7

As we noted, if c is a modular neighbor-distinguishing k-edge coloring of a graph G, then $c'(u) \neq c'(v)$ in $\mathbb{Z}_k$ for every pair u, v of adjacent vertices of G. Thus, the coloring c' is a proper vertex coloring of G with at most k colors, which implies that $\chi(G)$ is a lower bound for $\chi'_m(G)$.

Proposition 7.1. *For every connected graph G of order at least 3,*

$$\chi'_m(G) \geq \chi(G).$$

To illustrate these concepts, consider the tree T of order 10 in Fig. 7.1a. An edge coloring c of T is shown in Fig. 7.1b, where each edge is labeled with a color in $\mathbb{Z}_3 = \{0, 1, 2\}$ and each vertex is labeled with its induced color. Observe that $c'(u) \neq c'(v)$ in $\mathbb{Z}_3$ for every pair u, v of adjacent vertices of T. Thus, c is a modular 3-edge coloring of T and so $\chi'_m(T) \leq 3$. Since $\chi'_m(T) \geq 2$ by Proposition 7.1, it follows that $\chi'_m(T)$ is either 2 or 3. To show that $\chi'_m(T) = 3$, assume, to the contrary, that there exists a modular 2-edge coloring c of T. Thus $c'(v) = 0$ or $c'(v) = 1$ for each $v \in V(T)$. By the symmetry of the tree, we may assume that $c'(u_i) = 0$ and $c'(w_i) = 1$ for $1 \leq i \leq 5$. Hence, $c(u_iw_5) = 0$ and $c(u_5w_i) = 1$ for $1 \leq i \leq 4$. However, this implies that $c'(u_5) = c(u_5w_5) = c'(w_5)$, which is not possible. Therefore, $\chi'_m(T) \geq 3$ and so $\chi'_m(T) = 3 > \chi(T)$.

There are also graphs G for which $\chi'_m(G) = \chi(G)$. For example, consider the Petersen graph P in Fig. 7.2. Since $\chi(P) = 3$ and there exists a modular 3-edge coloring of P (also shown in the figure), $\chi'_m(P) = 3 = \chi(P)$.

a   **b**

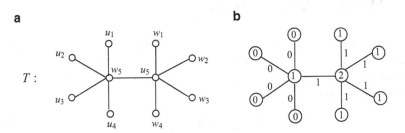

Fig. 7.1 A modular 3-edge coloring of a graph

Fig. 7.2 A modular 3-edge coloring of the Petersen graph

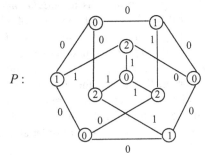

The following two results give the values of the modular chromatic indices of complete graphs and cycles.

Theorem 7.2 ([62]). *For each integer $n \geq 3$,*

$$\chi'_m(K_n) = \begin{cases} n+1 & \text{if } n \equiv 2 \pmod 4 \\ n & \text{otherwise.} \end{cases}$$

Theorem 7.3 ([62]). *For each integer $n \geq 3$,*

$$\chi'_m(C_n) = \begin{cases} 2 & \text{if } n \equiv 0 \pmod 4 \\ 3 & \text{if } n \not\equiv 0 \pmod 4. \end{cases}$$

By Theorems 7.2 and 7.3, there are two classes of connected graphs G of order $n \geq 3$ for which $\chi'_m(G) > \chi(G)$, namely the complete graphs K_n and the cycles C_n where $n \equiv 2 \pmod 4$. These are both special cases of a more general result.

Theorem 7.4 ([62]). *Let G be a connected graph of order 3 or more such that $\chi(G) \equiv 2 \pmod 4$. If each color class in every proper $\chi(G)$-coloring of G consists of an odd number of vertices, then $\chi'_m(G) > \chi(G)$.*

Proof. Since $\chi(G) \equiv 2 \pmod 4$, we can write $\chi(G) = 4p+2$ for some nonnegative integer p. Assume, to the contrary, that $\chi'_m(G) = \chi(G)$. Then there exists a modular $(4p+2)$-edge coloring $c : E(G) \rightarrow \mathbb{Z}_{4p+2}$ resulting in the vertex color classes $V_0, V_1, \ldots, V_{4p+1}$ where $c'(v) = i$ if $v \in V_i$ $(0 \leq i \leq 4p+1)$. If the vertex colors are summed, then each edge color is counted twice; that is,

$$\sum_{v \in V(G)} c'(v) = \sum_{i=0}^{4p+1} i \cdot |V_i| \equiv 2r \pmod{4p+2}$$

for some integer r with $0 \leq r \leq 2p$. However, this is impossible since each $|V_i|$ is odd and there is an odd number of odd integers i where $i \in \{0, 1, \ldots, 4p+1\}$. $\square$

7.2   Bipartite Graphs

In this section we determine the modular chromatic indices of bipartite graphs. This will play an important role in establishing the main result on this topic. We begin with an observation.

Observation 7.5. *Let G be a connected graph of order at least 3. If $c : E(G) \rightarrow \mathbb{N}$ is an edge coloring of G, then*

$$\sum_{v \in V(G)} c'(v) = 2 \sum_{e \in E(G)} c(e). \tag{7.1}$$

Thus if c is a modular k-edge coloring of G, then

$$\sum_{v \in V(G)} c'(v) \equiv 2 \sum_{e \in E(G)} c(e) \pmod{k}.$$

We first determine the modular chromatic index of a path. For integers $k \geq 2$ and $n \geq 3$ and an edge coloring c of the path $P_n = (v_1, v_2, \ldots, v_n)$, the *color sum sequence* of c is $s_c = (c'(v_1), c'(v_2), \ldots, c'(v_n))$, where $c'(v_i)$, $1 \leq i \leq n$, is the sum of the colors of the edges in P_n incident with v_i.

Theorem 7.6 ([62]). *For each integer $n \geq 3$,*

$$\chi'_m(P_n) = \begin{cases} 2 & \text{if } n \not\equiv 2 \pmod{4} \\ 3 & \text{if } n \equiv 2 \pmod{4}. \end{cases}$$

Proof. Let $P_n = (v_1, v_2, \ldots, v_n)$ where $n \geq 3$. For $n \equiv 0 \pmod{4}$ or $n \equiv 3 \pmod{4}$, define the 2-edge coloring $c_1 : E(P_n) \to \mathbb{Z}_2$ such that $c_1(v_i v_{i+1}) = 1$ if and only if $i \equiv 1, 2 \pmod{4}$. Then the color sum sequence $s_{c_1} = (c'_1(v_1), c'_1(v_2), \ldots, c'_1(v_n))$ of c_1 is

$$s_{c_1} = \begin{cases} (1, 0, 1, 0, \ldots, 1, 0) & \text{if } n \equiv 0 \pmod{4} \\ (1, 0, 1, 0, \ldots, 1, 0, 1) & \text{if } n \equiv 3 \pmod{4}. \end{cases}$$

For $n \equiv 1 \pmod{4}$, define the 2-edge coloring $c_2 : E(P_n) \to \mathbb{Z}_2$ such that $c_2(v_i v_{i+1}) = 1$ if and only if $i \equiv 2, 3 \pmod{4}$. Then the color sum sequence of c_2 is $s_{c_2} = (0, 1, 0, 1, 0, \ldots, 1, 0)$. Hence c_1 and c_2 are modular 2-edge colorings and so $\chi'_m(P_n) = 2$ if $n \not\equiv 2 \pmod{4}$.

For $n \equiv 2 \pmod{4}$, define the 3-edge coloring $c_3 : E(P_n) \to \mathbb{Z}_3$ by

$$c_3(v_i v_{i+1}) = \begin{cases} 0 & \text{if } i = n - 1 \\ 1 & \text{if } 1 \leq i \leq n - 2 \text{ and } i \equiv 1, 2 \pmod{4} \\ 2 & \text{if } 1 \leq i \leq n - 2 \text{ and } i \equiv 0, 3 \pmod{4}. \end{cases}$$

Then the color sum sequence of c_3 is

$$s_{c_3} = (1, 2, 0, 1, 0, 2, 0, 1, 0, \ldots, 2, 0, 1, 0, 2, 0, 1, 2, 0).$$

Thus c_3 is a modular 3-edge coloring and so $\chi'_m(P_n) \leq 3$. It then follows by Theorem 7.4 that $\chi'_m(P_n) = 3$. $\square$

Suppose next that G is a connected bipartite graph of order $n \geq 3$ with partite sets U and W where $|U| = r$ and $|W| = s$. If $\chi'_m(G) = 2$, then at least one of r and s must be even by Theorem 7.4. Next, we determine the modular chromatic indices of complete bipartite graphs.

Proposition 7.7 ([62]). *For positive integers r and s where $r + s \geq 3$,*

$$\chi'_m(K_{r,s}) = \begin{cases} 3 & \text{if r and s are odd} \\ 2 & \text{otherwise.} \end{cases}$$

Proof. We may assume that $1 \leq r \leq s$. First suppose that $r = 1$ and $s \geq 2$. If s is even, then the coloring that assigns the color 1 to every edge of $K_{1,s}$ is a modular 2-edge coloring. Hence, $\chi'_m(K_{1,s}) = 2$ in this case.

Suppose next that $r = 1$ and s is odd. By Theorem 7.4, $\chi'_m(K_{1,s}) \geq 3$. On the other hand, the coloring that assigns the color 1 of $\mathbb{Z}_3$ to two edges of $K_{1,s}$ and the color 0 to the remaining $s - 2$ edges is a modular 3-edge coloring of $K_{1,s}$. Thus the result holds for $r = 1$.

We now suppose that $r, s \geq 2$. By Proposition 7.1, $\chi'_m(K_{r,s}) \geq \chi(K_{r,s}) = 2$. Let U and W be the partite sets of $K_{r,s}$ with $|U| = r$ and $|W| = s$. If at least one of r and s, say r, is even, then let $w \in W$ and consider a 2-edge coloring that assigns the color 1 to an edge e if and only if e is incident with w. Then this is a modular 2-edge coloring of $K_{r,s}$ and so $\chi'_m(K_{r,s}) = 2$.

If both r and s are odd, then $\chi'_m(K_{r,s}) \geq 3$ by Theorem 7.4. Thus, $r = 6p + q \geq 3$, where p is a nonnegative integer and $q \in \{1, 3, 5\}$. Let $w \in W$. If $q \neq 1$, then the edge coloring c_1 in which $c_1(e) = 1$ if e is incident with w and $c_1(e) = 0$ otherwise is a modular 3-edge coloring of $K_{r,s}$. If $q = 1$, then $r \geq 7$. Let $U = \{u_1, u_2, \ldots, u_r\}$ and observe that the edge coloring c_2 given by

$$c_2(e) = \begin{cases} 2 & \text{if } e \in \{u_1 w, u_2 w\} \\ 1 & \text{if } e = u_i w \ (3 \leq i \leq r) \\ 0 & \text{otherwise} \end{cases}$$

is a modular 3-edge coloring of $K_{r,s}$. ☐

Since every tree T of order at least 3 is bipartite, it follows by Theorem 7.4 that $\chi'_m(T) \geq 2$. In fact, it was shown in [62] that for every tree T of order at least 3, either $\chi'_m(T) = 2$ or $\chi'_m(T) = 3$. Indeed, it is known exactly which trees have exactly each of these two values.

Theorem 7.8 ([62]). *If T is a tree of order $r + s \geq 3$ whose partite sets have orders r and s, then*

$$\chi'_m(T) = \begin{cases} 3 & \text{if r and s are odd} \\ 2 & \text{otherwise.} \end{cases}$$

Proof. We first show that every nontrivial tree of odd order is modular 2-edge colorable. Assume, to the contrary, that there exists a tree of odd order whose modular chromatic index is greater than 2. Let T be such a tree of minimum order $r + s$ and suppose that U and W are the partite sets of T with $|U| = r$ and $|W| = s$.

It follows by Proposition 7.7 that T is not a star and so we may assume that $r+s \geq 5$ where $r \geq 2$ is even and $s \geq 3$ is odd. Also, since T is not a path by Theorem 7.6, there are at least three end-vertices, implying that there are two end-vertices x and y belonging to the same partite set. Let T_1 be the tree obtained from T by deleting x and y. Therefore, $\chi'_m(T_1) = 2$ by assumption.

Let c_1 be a modular 2-edge coloring of T_1. Furthermore, let $U_1 \subseteq U$ and $W_1 \subseteq W$ be the partite sets of T_1 and observe that $|U_1|$ is even while $|W_1|$ is odd. Hence, c_1 assigns colors to the edges of T_1 so that $c'_1(v) = 1$ if and only if $v \in U_1$ by Observation 7.5. If $x, y \in W$, then the edge coloring c of T given by $c(e) = c_1(e)$ if $e \in E(T_1)$ and $c(e) = 0$ otherwise is a modular 2-edge coloring of T, which contradicts our assumption. Thus, we may assume that $x, y \in U$.

Let $w_1 \in N(x)$ and $w_2 \in N(y)$ and consider the $w_1 - w_2$ path P in T_1. (If $d(x, y) = 2$, then $w_1 = w_2$ and so $E(P) = \emptyset$.) We define an edge coloring c of T as follows:

$$c(e) = \begin{cases} c_1(e) + 1 & \text{if } e \in E(P) \\ 1 & \text{if } e \in \{xw_1, yw_2\} \\ c_1(e) & \text{otherwise.} \end{cases}$$

We verify that c is a modular 2-edge coloring of T. If $v \in V(T_1) - V(P)$, then $c'(v) = c'_1(v)$; while if $v \in V(P)$, then $c'(v) = c'_1(v) + 2 = c'_1(v)$. Hence, $c'(v) = c'_1(v)$ for every $v \in V(T_1)$; that is, $c'(v) = 1$ if $v \in U_1$ and $c'(v) = 0$ if $v \in W_1$. Since $c'(x) = c'(y) = 1$, this is a modular 2-edge coloring of T, which again is impossible. Hence, such a tree T does not exist and so $\chi'_m(T) = 2$ if $r + s$ is odd.

Next, assume that $r + s \geq 4$ is even. If both r and s are even, then it can be verified that T is modular 2-edge colorable by an argument similar to that used in the case when $r + s$ is odd. Thus we may assume that both r and s are odd. Let $r + s = 2k$ where $k \geq 2$. We need only verify that $\chi'_m(T) \leq 3$ by Theorem 7.4. We proceed by induction on k. For $k = 2$, $T = K_{1,3}$ and the result immediately follows by Proposition 7.7. Suppose for some integer $k \geq 2$ that every tree of order $2k$ which is a spanning subgraph of $K_{r,2k-r}$ for some odd integer r ($1 \leq r \leq 2k - 1$) is modular 3-edge colorable.

Let T be a tree of order $2(k+1)$ with $T \subseteq K_{r,2(k+1)-r}$ for some odd integer r with $1 \leq r \leq 2(k+1) - 1$. Since T is not a star, let U and W be the partite sets of T such that $|U| = r \geq 3$ and $|W| = 2(k+1) - r \geq 3$. Also, since T is not a path, there exist at least three end-vertices in T, two of which belong to the same partite set. We may assume that x and y are end-vertices, both belonging to U. Also, let w_1 and w_2 be the vertices in W such that $xw_1, yw_2 \in E(T)$. Consider the tree T_1 of order $2k$ obtained from T by deleting x and y. Then the sets $U_1 = U - \{x, y\}$ and $W_1 = W$ are the partite sets of T_1 and, furthermore, both $|U_1|$ and $|W_1|$ are odd. Hence, $\chi'_m(T_1) = 3$ and so let $c_1 : E(T_1) \rightarrow \mathbb{Z}_3$ be a modular 3-edge coloring of T_1. We consider three cases.

Case 1. $0 \in \{c_1'(v) : v \in U_1\}$.   Then the edge coloring c given by $c(xw_1) = c(yw_2) = 0$ and $c(e) = c_1(e)$ for every $e \in E(T_1)$ is a modular 3-edge coloring of T.

Case 2. $\{c_1'(v) : v \in W_1\} = \{0\}$.   Note that $d(x, y) = d$ is a positive even integer. Let $P = (w_1 = v_1, v_2, \ldots, v_{d-1} = w_2)$ be the $w_1 - w_2$ path in T_1. (If $d = 2$, then $w_1 = w_2$ and $E(P) = \emptyset$.) Therefore, $v_i \in W_1$ if i is odd and $v_i \in U_1$ if i is even. Define an edge coloring c of T by

$$
c(e) = \begin{cases}
c_1(e) + 1 & \text{if } e = v_i v_{i+1} \in E(P) \text{ and } i \text{ is odd} \\
c_1(e) + 2 & \text{if } e = v_i v_{i+1} \in E(P) \text{ and } i \text{ is even} \\
2 & \text{if } e = xw_1 \\
1 & \text{if } e = yw_2 \\
c_1(e) & \text{otherwise.}
\end{cases}
$$

To verify that c is a modular 3-edge coloring of T, first observe that $c'(v) = c_1'(v)$ for every $v \in V(T_1) - V(P)$. Also, $c'(v) = c_1'(v) + 3 = c_1'(v)$ for every $v \in V(P)$. In particular, $c'(w_1) = c'(w_2) = 0$. Thus, $c'(x) = 1 \neq c'(w_1)$ and $c'(y) = 2 \neq c'(w_2)$, implying that c is a modular 3-edge coloring of T.

Case 3. $\{c_1'(v) : v \in U_1\} = \{a\}$ and $b \in \{c_1'(v) : v \in W_1\}$ where $\{a, b\} = \{1, 2\}$. We consider three subcases.

Subcase 3.1. $d(x, y) = d \geq 4$.   If $c_1'(w_1) = c_1'(w_2) = b$, then let c be an edge coloring of T such that $c(xw_1) = c(yw_2) = a$ and $c(e) = c_1(e)$ for every $e \in E(T_1)$ and observe that c is a modular 3-edge coloring of T_1.

If $c_1'(w_1) = 0$ or $c_1'(w_2) = 0$, say the former, then let $P = (w_1 = v_1, v_2, \ldots, v_{d-1} = w_2)$ be the $w_1 - w_2$ path in T_1 and define an edge coloring c of T by

$$
c(e) = \begin{cases}
c_1(e) + a & \text{if } e = v_i v_{i+1} \in E(P) \text{ and } i \text{ is odd} \\
c_1(e) + b & \text{if } e = v_i v_{i+1} \in E(P) \text{ and } i \text{ is even} \\
a & \text{if } e \in \{xw_1, yw_2\} \\
c_1(e) & \text{otherwise.}
\end{cases}
$$

Then

$$
c'(v) = \begin{cases}
a & \text{if } v \in \{x, y\} \\
2a = b & \text{if } v = w_1 \\
c_1'(v) & \text{otherwise}
\end{cases}
$$

and it is straightforward to verify that this is a modular 3-edge coloring of T.

Subcase 3.2. $d(x, y) = 2$.   Let $w_1 = w_2 = w$. If $c_1'(w) = 0$, then let c be an edge coloring such that $c(xw) = c(yw) = a$ and $c(e) = c_1(e)$ for every $e \in E(T_1)$ and observe that this is a modular 3-edge coloring of T.

Hence, suppose finally that $c_1'(w) = b$. Since T is not a star, there exists an end-vertex z in T such that $d(x, z) \geq 3$. Let $P = (w = v_1, v_2, \ldots, v_d = z)$ be the $w - z$ path in T_1, where $d = d(x, z)$.

Subcase 3.2.1. *d is odd.*   Then $z \in W$ and so $c_1'(z) \in \{0, b\}$. Then the edge coloring c defined by

$$c(e) = \begin{cases} c_1(e) - c_1'(z) + a & \text{if } e = v_i v_{i+1} \in E(P) \text{ and } i \text{ is odd} \\ c_1(e) + c_1'(z) + b & \text{if } e = v_i v_{i+1} \in E(P) \text{ and } i \text{ is even} \\ a & \text{if } e \in \{xw, yw\} \\ c_1(e) & \text{otherwise} \end{cases}$$

is a modular 3-edge coloring of T because $c_1'(z) \in \{0, b\}$ and

$$c'(v) = \begin{cases} a & \text{if } v \in \{x, y\} \\ 2c_1'(z) + b \in \{0, b\} & \text{if } v = z \\ b - c_1'(z) \in \{0, b\} & \text{if } v = w \\ c_1'(v) & \text{otherwise.} \end{cases}$$

Subcase 3.2.2. *d is even.*   Then $z \in U$ and so $c_1'(z) = a$. Let w_3 be the neighbor of z in T, that is, $w_3 = v_{d-1}$. Then consider the edge coloring c defined by

$$c(e) = \begin{cases} c_1(e) - c_1'(w_3) + a & \text{if } e = v_i v_{i+1} \in E(P) \text{ and } i \text{ is odd and } i \neq d - 1 \\ c_1(e) + c_1'(w_3) + b & \text{if } e = v_i v_{i+1} \in E(P) \text{ and } i \text{ is even} \\ a & \text{if } e \in \{xw, yw, zw_3\} \\ c_1(e) & \text{otherwise.} \end{cases}$$

It can be verified that c is a modular 3-edge coloring of T. $\square$

With the aid of Theorem 7.8, it is possible to determine the modular chromatic index of every connected bipartite graph of order at least 3.

Theorem 7.9 ([62]). *If G is a connected bipartite graph of order $r + s \geq 3$ such that $G \subseteq K_{r,s}$, then*

$$\chi_m'(G) = \begin{cases} 3 & \text{if } r \text{ and } s \text{ are odd} \\ 2 & \text{otherwise.} \end{cases}$$

Proof. If G is a tree, then the result holds by Theorem 7.8. If G is not a tree, then $T \subseteq K_{r,s}$ for each spanning tree T of G. Let c_T be a modular edge coloring of T and define an edge coloring c of G by $c(e) = c_T(e)$ if $e \in E(T)$ and $c(e) = 0$ otherwise. Then $c'(v) = c'_T(v)$ for every vertex v in G. Therefore, every modular edge coloring of T induces a modular edge coloring of G using the same number of colors, which implies that $\chi'_m(G) \leq \chi'_m(T)$. The result follows by Theorems 7.4 and 7.8. □

7.3   Modular Chromatic Index and Chromatic Number

For every graph G that has been encountered thus far, either $\chi'_m(G) = \chi(G)$ or $\chi'_m(G) = \chi(G) + 1$. This is no coincidence. In order to verify this fact, we first present another theorem. It will be convenient to have an additional definition. If g and h are two vertex colorings of a graph G such that $g : V(G) \to \{1, 2, \ldots, r\} = [r]$ and $h : V(G) \to \mathbb{Z}_r$ for some integer $r \geq 2$ such that $g(v) = i$ and $h(v) = i \in \mathbb{Z}_r$ for every vertex v of G, then we say $g = h$.

Theorem 7.10 ([63]). *Let G be a connected graph of order at least 3 that is not bipartite. If G is $(2k + 1)$-colorable where $k \in \mathbb{N}$, then G is modular $(2k + 1)$-edge colorable. Furthermore, for a given proper $(2k + 1)$-vertex coloring $f : V(G) \to \{1, 2, \ldots, 2k + 1\}$, there is a modular $(2k + 1)$-edge coloring $c : E(G) \to \mathbb{Z}_{2k+1}$ such that $c' = f$.*

Proof. Let $V(G) = \{v_1, v_2, \ldots, v_n\}$, $n \geq 3$, and let $f : V(G) \to \{1, 2, \ldots, 2k + 1\}$ be a proper $(2k + 1)$-vertex coloring of G. A sequence of $n + 1$ edge colorings $c_0, c_1, \ldots, c_n$, where $c_i : E(G) \to \mathbb{Z}_{2k+1}$ for $0 \leq i \leq n$, is defined recursively by (i) if $i = 0$, then $c'_0(v) = 0$ for $v \in V(G)$ and (ii) if $1 \leq i \leq n$, then $c'_i(v_j) = c'(v_j)$ for $1 \leq j \leq i$ and $c'_i(v) = 0$ for $v \in V(G) - \{v_1, v_2, \ldots, v_i\}$. This will imply that $c = c_n$ is a modular $(2k + 1)$-edge coloring with $c'(v) = f(v)$ for every $v \in V(G)$.

First, we define the edge coloring $c_0 : E(G) \to \mathbb{Z}_{2k+1}$ by $c_0(e) = 0$ for all $e \in E(G)$. Thus $c'_0(v) = 0$ for $v \in V(G)$. Next, we define the edge coloring $c_1 : E(G) \to \mathbb{Z}_{2k+1}$ of G from c_0 such that $c'_1(v_1) = c'(v_1)$ and $c'_1(v) = 0$ if $v \in V(G) - \{v_1\}$. Suppose that $f(v_1) = a$. Since $\gcd(2, 2k + 1) = 1$, it follows that $2 \mid a$ in $\mathbb{Z}_{2k+1}$ and so $a = 2b$ for some $b \in \mathbb{Z}_{2k+1}$. We consider two cases.

Case 1. v_1 lies on an odd cycle C of G. Let

$$C = (v_1 = u_1, u_2, \ldots, u_p, u_{p+1} = u_1)$$

where $p \geq 3$ is an odd integer. The coloring $c_1 : E(G) \to \mathbb{Z}_{2k+1}$ is defined by

$$c_1(e) = \begin{cases} c_0(e) & \text{if } e \notin E(C) \\ c_0(e) + b & \text{if } e = u_i u_{i+1}, i \text{ is odd and } 1 \leq i \leq p \\ c_0(e) - b & \text{if } e = u_i u_{i+1}, i \text{ is even and } 2 \leq i \leq p - 1. \end{cases} \quad (7.2)$$

Thus $c'_1(v_1) = 2b = a = f(v_1)$ and $c'_1(v_i) = c'_0(v_i) = 0$ for $2 \leq i \leq n$.

Case 2. v_1 *lies on no odd cycle of* G. Since G is connected, there is a path P joining v_1 and a vertex on an odd cycle C of G. Suppose that

$$C = (u_1, u_2, \ldots, u_p, u_{p+1} = u_1)$$

in G. We may assume, without loss of generality that

$$P = (v_1 = w_1, w_2, \ldots, w_t = u_1)$$

such that u_1 is the only vertex on P that belongs to C. We consider two subcases, according to whether t is even or t is odd.

Subcase 2.1. t *is even.* The coloring $c_1 : E(G) \to \mathbb{Z}_{2k+1}$ is defined by

$$c_1(e) = \begin{cases} c_0(e) & \text{if } e \notin E(C) \cup E(P) \\ c_0(e) + a & \text{if } e = w_j w_{j+1}, j \text{ is odd and } 1 \le j \le t-1 \\ c_0(e) - a & \text{if } e = w_j w_{j+1}, j \text{ is even and } 2 \le j \le t-2 \\ c_0(e) - b & \text{if } e = u_i u_{i+1}, i \text{ is odd and } 1 \le i \le p \\ c_0(e) + b & \text{if } e = u_i u_{i+1}, i \text{ is even and } 2 \le i \le p-1. \end{cases} \qquad (7.3)$$

Thus $c_1'(v_1) = 2b = a = f(v_1)$ and $c_1'(v_i) = c_0'(v_i) = 0$ for $2 \le i \le n$. Note that $c_1'(u_1) = a + (-2b) = 0$.

Subcase 2.2. t *is odd.* The coloring $c_1 : E(G) \to \mathbb{Z}_{2k+1}$ is defined by

$$c_1(e) = \begin{cases} c_0(e) & \text{if } e \notin E(C) \cup E(P) \\ c_0(e) + a & \text{if } e = w_j w_{j+1}, j \text{ is odd and } 1 \le j \le t-1 \\ c_0(e) - a & \text{if } e = w_j w_{j+1}, j \text{ is even and } 2 \le j \le t-2 \\ c_0(e) + b & \text{if } e = u_i u_{i+1}, i \text{ is odd and } 1 \le i \le p \\ c_0(e) - b & \text{if } e = u_i u_{i+1}, i \text{ is even and } 2 \le i \le p-1. \end{cases} \qquad (7.4)$$

Thus $c_1'(v_1) = 2b = a = f(v_1)$ and $c_1'(v_i) = c_0'(v_i) = 0$ for $2 \le i \le n$. Note that $c_1'(u_1) = -a + 2b = 0$.

In each case, $c_1'(v_1) = c'(v_1)$ and $c_1'(v) = c_0'(v) = 0$ for all $v \in V(G) - \{v_1\}$. (The coloring c_1 is neither a proper edge coloring nor a modular edge coloring of G.) In general, for an integer i with $1 \le i \le n-1$, suppose that the coloring $c_i : E(G) \to \mathbb{Z}_{2k+1}$ is defined such that $c_i'(v_j) = f(v_j)$ for $1 \le j \le i$ and $c_i'(v) = 0$ for all $v \in V(G) - \{v_1, v_2, \ldots, v_i\}$. Then the coloring $c_{i+1} : E(G) \to \mathbb{Z}_{2k+1}$ is defined from c_i in the same fashion as described in (7.2)–(7.3), namely by replacing c_0 and c_1 in (7.2)–(7.3) by c_i and c_{i+1}, respectively. An argument similar to the one used in the case dealing with c_1

and c_0 shows that $c'_{i+1}(v_j) = c_i'(v_j) = f(v_j)$ for $1 \le j \le i$, $c'_{i+1}(v_{i+1}) = f(v_{i+1})$ and $c'_{i+1}(v) = 0$ for $v \in V(G) - \{v_1, v_2, \ldots, v_{i+1}\}$. In particular, $c_n : E(G) \to \mathbb{Z}_{2k+1}$ has the property that $c'_n(v_i) = c'_{n-1}(v_i) = f(v_i)$ for $1 \le i \le n - 1$ and $c'_n(v_n) = f(v_n)$. Therefore, c_n is a modular $(2k + 1)$-edge coloring of G and so G is modular $(2k + 1)$-edge colorable. $\square$

The following corollaries are consequences of Proposition 7.1 and Theorem 7.10.

Corollary 7.11 ([63]). *If G is a connected graph of order at least 3, then*

$$\chi(G) \le \chi'_m(G) \le \chi(G) + 1.$$

Furthermore, if $\chi(G)$ is odd, then $\chi'_m(G) = \chi(G)$.

Proof. We have seen that $\chi'_m(G) \ge \chi(G)$ in Proposition 7.1. For the upper bound, let G be a connected graph of order $n \ge 3$. If $G = K_n$, then the result follows by Theorem 7.2. Thus, we may assume that $G \ne K_n$ and $\chi(G) = k \le n - 1$. If k is even, then G is $(k + 1)$-colorable and so G is modular $(k + 1)$-edge colorable by Theorem 7.10. Thus $\chi'_m(G) \le \chi(G) + 1$. If k is odd, then G is k-colorable and so G is modular k-edge colorable by Theorem 7.10. Therefore, $\chi'_m(G) \le \chi(G)$ and so $\chi'_m(G) = \chi(G)$. $\square$

By Corollary 7.11, if G is a connected graph of order at least 3 such that $\chi'_m(G) = \chi(G) + 1$, then $\chi(G)$ is even and so $\chi(G) \equiv 0 \pmod 4$ or $\chi(G) \equiv 2 \pmod 4$. By Theorem 7.4 and Corollary 7.11, we have the following result for connected graphs G with $\chi(G) \equiv 2 \pmod 4$.

Corollary 7.12 ([63]). *Suppose that G is a connected graph of order at least 3 such that $\chi(G) \equiv 2 \pmod 4$. If each color class in every proper $\chi(G)$-coloring of G consists of an odd number of vertices, then $\chi'_m(G) = \chi(G) + 1$.*

The following theorem gives a necessary and sufficient condition for the modular chromatic index of a graph G to equal $\chi(G) + 1$.

Theorem 7.13 ([63]). *Let G be a connected graph of order at least 3. Then $\chi'_m(G) = \chi(G) + 1$ if and only if $\chi(G) \equiv 2 \pmod 4$ and every proper $\chi(G)$-coloring of G results in color classes of odd size.*

Chapter 8
Strong Edge Colorings of Graphs

In the preceding chapters we have discussed unrestricted edge colorings that, in a variety of ways, induce vertex colorings that are either vertex-distinguishing or neighbor-distinguishing. In this chapter, we turn our attention from unrestricted edge colorings to proper edge colorings that induce set-defined vertex colorings which are either vertex-distinguishing or neighbor-distinguishing. Proper edge colorings inducing set-defined vertex colorings have been referred to as strong edge colorings. We first discuss vertex-distinguishing strong edge colorings, which, typically, have been called simply strong edge colorings.

8.1   The Strong Chromatic Index

In 1997 Burris and Schelp [18] defined a *strong edge coloring* of G as a proper edge coloring that induces a vertex-distinguishing coloring which assigns to each vertex v of G the set $S(v)$ of colors of the edges incident with v. Since the edge coloring is vertex-distinguishing, no two vertices of G are colored the same, that is, for every two vertices of G, there is some color that is assigned to an edge incident with one of these two vertices that is not assigned to any edge incident with the other vertex. The minimum positive integer k for which G has a strong k-edge coloring is called the *strong chromatic index* of G and is denoted by $\chi'_s(G)$. Since every strong edge coloring of a nonempty graph G is a proper edge coloring of G, it follows that

$$\Delta(G) \leq \chi'(G) \leq \chi'_s(G).$$

As an example, we determine the strong chromatic index of the graph G of Fig. 8.1a. Since $\chi'(G) = 3$, it follows that $\chi'_s(G) \geq 3$. However, $\chi'_s(G) \neq 3$, since any proper 3-edge coloring of G would assign the color $\{1, 2, 3\}$ to every vertex of degree 3. Moreover, if $\chi'_s(G) = 3$, then since the order of G is 7 the seven vertices

© Ping Zhang 2015
P. Zhang, *Color-Induced Graph Colorings*, SpringerBriefs in Mathematics,
DOI 10.1007/978-3-319-20394-2_8

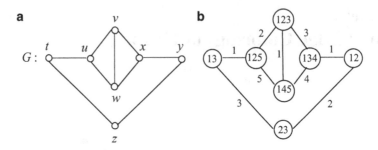

Fig. 8.1 A strong 5-edge coloring of a graph

of G would have to be colored with the seven nonempty sets of $\{1,2,3\}$. Since $\delta(G) = 2$, no vertex of G can be colored $\{1\}, \{2\}$ or $\{3\}$. Furthermore, $\chi'_s(G) \neq 4$, for suppose that there is a strong 4-edge coloring c of G. We may assume that $c(vw) = 1$. Since c is a proper edge coloring, none of the edges uv, uw, vx and wx can be colored 1. Hence, two of these four edges must be assigned the same color and the remaining two edges must be assigned different colors, say uv and wx are colored 2. Thus, all of the vertices u, v, w and x are assigned a color that is a 3-element set containing 2. This, however, implies that two of these vertices are colored the same, which is impossible. Hence, $\chi'_s(G) \geq 5$. The strong 5-edge coloring of G in Fig. 8.1b shows that $\chi'_s(G) = 5$.

While the graph G of order 7 in Fig. 8.1 does not have strong chromatic index 3, there are connected graphs of order 7 that do. Suppose that H is a connected graph of order 7 with $\chi'_s(H) = 3$. Then there exists a strong 3-edge coloring c of G using the colors 1, 2, 3. Since c is vertex-distinguishing, the colors of the seven vertices of H must be the seven nonempty sets of $\{1,2,3\}$. Because three of these subsets have one element, three have two elements and the other has three elements, the degree sequence of H must be 3, 2, 2, 2, 1, 1, 1 and so H has size 6, which implies that H is a tree. There are three trees with this degree sequence and all have strong chromatic index 3 (see Fig. 8.2).

The argument used to verify that the strong chromatic index of the graph G of Fig. 8.1a is 5 suggests a more general observation. If a connected graph G has strong chromatic index k, say, then the induced color assigned to a vertex of degree r is one of the r-element subsets of $\{1, 2, \ldots, k\}$.

Observation 8.1. *If G is a connected graph of order at least 3 containing more than $\binom{k}{r}$ vertices of degree r $(1 \leq r \leq \Delta(G))$ for some positive integer k, then*

$$\chi'_s(G) \geq k + 1.$$

An another illustration, suppose that G is a connected graph of order n with $\chi'_s(G) = k$. Then $n \leq 2^k - 1$. Furthermore, if $\chi'_s(G) = k$ and G has order $2^k - 1$, then G must contain exactly $\binom{k}{r}$ vertices of degree r for every integer r with $1 \leq r \leq k$. For example, the graph G of order $15 = 2^4 - 1$ in Fig. 8.3 has $\binom{4}{r}$ vertices of degree r

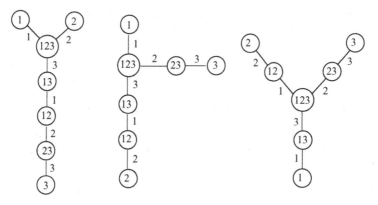

Fig. 8.2 The trees of order 7 with strong chromatic index 3

Fig. 8.3 A graph of order 15 with strong chromatic index 4

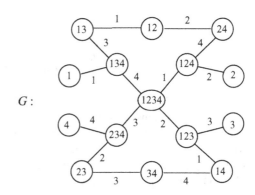

for every integer r with $1 \leq r \leq 4$. The proper 4-edge coloring of G in Fig. 8.3 has the property that $\{S(v) : v \in V(G)\} = \mathscr{P}([4]) - \{\emptyset\}$ and so $\chi_s'(G) = 4$.

Although $\Delta(G) + 1$ is an upper bound for $\chi'(G)$ by Vizing's Theorem, $\Delta(G) + 1$ is not an upper bound for $\chi_s'(G)$. In fact, there is no constant c such that $\chi_s'(G) \leq \Delta(G) + c$ for every graph G since, for example, if $n = \binom{\ell+1}{2} + 1$, then $\chi_s'(C_n) \geq \ell + 2 = \Delta(C_n) + \ell$ by Observation 8.1. However, for a connected graph G of order n with $\Delta(G) \geq 2$, the number $n + \Delta(G) - 1$ *is* an upper bound for $\chi_s'(G)$.

Theorem 8.2. *If G is a connected graph of order $n \geq 3$, then*

$$\chi_s'(G) \leq n + \Delta(G) - 1.$$

Proof. By Vizing's theorem, there exists a proper $(\Delta(G) + 1)$-edge coloring of G. Let such an edge coloring of G be given using the colors $1, 2, \ldots, \Delta(G) + 1$. Also, let $v_1, v_2, \ldots, v_n$ be an ordering of the vertices of G such that for each i ($2 \leq i \leq n$), the vertex v_i is adjacent to one or more of the vertices preceding it in the ordering. For each i with $2 \leq i \leq n - 1$, select an edge joining v_i and some vertex in $\{v_1, v_2, \ldots, v_{i-1}\}$ and replace the color of this edge by $\Delta(G) + i$. Now,

$$S(v_n) \subseteq \{1, 2, \ldots, \Delta(G) + 1\}$$

and, for i and j with $2 \leq i < j \leq n-1$, the vertex v_i is incident with an edge colored $\Delta(G) + i$ while v_j is not. Hence, $S(v_i) \neq S(v_j)$ for all pairs i, j of distinct integers with $2 \leq i, j \leq n$. Necessarily, the only edge assigned the color $\Delta(G) + 2$ is $v_2 v_1$. Consequently, $S(v_1) \neq S(v_i)$ for $3 \leq i \leq n$. Since exactly one of the edges $v_3 v_2$ and $v_3 v_1$ is assigned the color $\Delta(G) + 3$, it follows that $S(v_1) \neq S(v_2)$. Hence, the vertices of G are assigned distinct labels and so the resulting $(n + \Delta(G) - 1)$-edge coloring of G is a strong edge coloring. Therefore, $\chi'_s(G) \leq n + \Delta(G) - 1$. $\square$

A much improved bound for the strong chromatic index was obtained by Bazgan, Harkat-Benhamdine, Li and Woźniak, verifying a conjecture by Burris and Schelp (see [12]).

Theorem 8.3. *If G is a connected graph of order $n \geq 3$, then $\chi'_s(G) \leq n + 1$.*

That the bound in Theorem 8.3 is sharp is illustrated by the next result.

Theorem 8.4. *If $n \geq 3$, then*

$$\chi'_s(K_n) = \begin{cases} n & \text{if } n \text{ is odd} \\ n + 1 & \text{if } n \text{ is even.} \end{cases}$$

Proof. Since it's readily observed that $\chi'_s(K_3) = 3$, we may assume that $n \geq 4$. Because K_n contains n vertices of degree $n - 1$, it follows by Observation 8.1 that $\chi'_s(K_n) \geq n$. We write $n = 2k$ if n is even and $n = 2k + 1$ if n is odd, where $k \geq 2$ in each case. First, consider the complete graph K_{2k+2} with vertex set $\{v_0, v_1, \ldots, v_{2k+1}\}$. Place the vertices $v_1, v_2, \ldots, v_{2k+1}$ cyclically about a regular $(2k + 1)$-gon and place v_0 in the center of the $(2k + 1)$-gon. Join every two vertices of K_{2k+2} by a straight line segment. For each integer i with $1 \leq i \leq 2k + 1$, the edge $v_0 v_i$ and all edges perpendicular to $v_0 v_i$ form a 1-factor F_i of K_{2k+2} and so

$$\mathscr{F} = \{F_1, F_2, \ldots, F_{2k+1}\}$$

is a 1-factorization of K_{2k+2}. Assign each edge of F_i the color i for $1 \leq i \leq 2k + 1$. By deleting v_1 from K_{2k+2}, we obtain K_{2k+1} with vertex set

$$\{v_0, v_2, v_3, \ldots, v_{2k+1}\}.$$

Since the edges $v_1 v_i$ with $i \in \{0, 2, 3, \ldots, 2k + 1\}$ have $2k + 1$ distinct colors, the induced vertex labels for the vertices of K_{2k+1} are distinct and so the $(2k + 1)$-edge coloring of K_{2k+1} is a strong edge coloring. Thus, $\chi'_s(K_n) \leq 2k + 1$. Since $\chi'_s(K_{2k+1}) \geq \chi'(K_{2k+1}) = 2k + 1$, it follows that $\chi'_s(K_n) = n$ if n is odd.

By deleting both v_1 and v_2 from K_{2k+2}, we obtain K_{2k} with vertex set

$$\{v_0, v_3, v_4, \ldots, v_{2k+1}\}.$$

Let $S = \{1, 2, \ldots, 2k + 1\}$. Since the colors of the edges incident with v_0 in K_{2k} are those in $S - \{1, 2\}$, the colors of the edges incident with v_{2i+1} ($1 \leq i \leq k$) are those in $S - \{i + 1, k + i + 2\}$ and the colors of the edges incident with v_{2i+2} ($1 \leq i \leq k - 1$) are those in $S - \{i + 2, k + i + 2\}$, where each color is one of $1, 2, \ldots, 2k + 1$, modulo $2k + 1$, the $(2k + 1)$-edge coloring of K_{2k} is a strong edge coloring. Thus, $\chi'_s(K_{2k}) \leq 2k + 1$.

It remains therefore only to show that $\chi'_s(K_{2k}) = 2k + 1$. Suppose, to the contrary, that $\chi'_s(K_{2k}) = 2k$. Since K_{2k} contains $2k$ vertices of degree $2k - 1$ and $\binom{2k}{2k-1} = 2k$, each possible $(2k - 1)$-element subset of $\{1, 2, \ldots, 2k\}$ must be the label of some vertex of K_{2k}. This implies that for each $i \in \{1, 2, \ldots, 2k\}$, there is exactly one vertex of K_{2k} not incident with an edge colored i. Suppose that v_{2k} is the only vertex of K_{2k} not incident with an edge colored 1. Then each of $v_1, v_2, \ldots, v_{2k-1}$ is incident with an edge colored 1. Hence, at least one of these vertices, say v_1, is incident with at least two edges colored 1. This, however, implies that the number of colors of the edges incident with v_1 is at most $2k - 2$, which is impossible. Therefore, $\chi'_s(K_n) = n + 1$ if n is even. □

Accordingly, there is similarity in the formulas for the chromatic index and the strong chromatic index of complete graphs, namely

$$\chi'(K_n) = 2\left\lceil \frac{n}{2} \right\rceil - 1 \text{ and } \chi'_s(K_n) = 2\left\lfloor \frac{n}{2} \right\rfloor + 1.$$

8.2 Binomial Colorings of Graphs

We saw in the preceding section that a strong k-edge coloring of a graph G is a set-induced, vertex-distinguishing proper edge coloring $c : E(G) \rightarrow [k] = \{1, 2, \ldots, k\}$, $k \geq 2$. Therefore, adjacent edges are assigned distinct colors and c induces the vertex coloring $c' : V(G) \rightarrow \mathscr{P}([k])$ defined so that $c'(v)$ is the set of colors of the edges incident with v, where $\mathscr{P}([k])$ is the power set of $[k]$. If G is a graph of order n with strong chromatic index $\chi'_s(G) = k$, then $n \leq 2^k = |\mathscr{P}([k])|$. Furthermore, if $\chi'_s(G) = k$ and G has order 2^k, then G must contain exactly $\binom{k}{r}$ vertices of degree r for every integer r with $0 \leq r \leq k$. For a given integer $k \geq 2$, we seek the largest order of a graph possessing a strong k-edge coloring. This suggests the study of a class of graphs, which was introduced in [21].

For an integer $k \geq 2$, a k-*binomial graph* is a graph containing $\binom{k}{r}$ vertices of degree r for each integer r with $0 \leq r \leq k$. Thus, such a graph G has order $n = \sum_{r=0}^{k} \binom{k}{r} = 2^k$ and size

$$m = \frac{1}{2} \sum_{r=0}^{k} r \binom{k}{r} = k2^{k-2}.$$

Fig. 8.4 The k-binomial
graphs for $k = 2, 3$

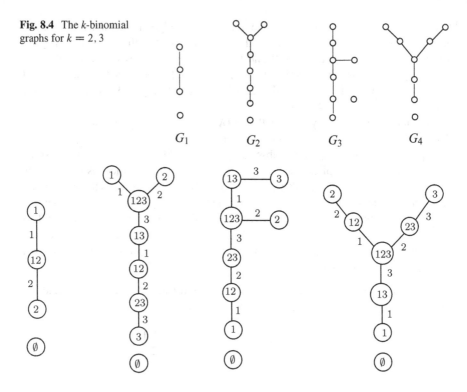

G_1 G_2 G_3 G_4

Fig. 8.5 Four proper binomial-colored graphs

A graph G is a *binomial graph* if G is a k-binomial graph for some integer $k \geq 2$.
There is only one 2-binomial graph and three 3-binomial forests. All four of these
graphs are shown in Fig. 8.4.

For an integer $k \geq 2$, a *proper k-binomial-colored graph* G is a graph with a
proper edge coloring $c : E(G) \rightarrow [k] = \{1, 2, \ldots, k\}$ such that the induced vertex
coloring $c' : V(G) \rightarrow \mathscr{P}([k])$, where $c'(v)$ is the set of colors of the edges incident
with v, is vertex-distinguishing and satisfies the condition that

$$\{c'(v) : v \in V(G)\} = \mathscr{P}([k]).$$

In this case, the edge coloring c is referred to as a *proper k-binomial-coloring* of G.
A graph G is a *proper binomial-colored graph* if G is a proper k-binomial-colored
graph for some integer $k \geq 2$. There concepts were introduced and studied in [21].
Necessarily, a proper binomial-colored graph is a binomial graph. Each graph in
Fig. 8.4 is a proper binomial-colored graph as is shown in Fig. 8.5. Furthermore, the
graph of Fig. 8.3 is a proper 4-binomial-colored graph. Note that for each integer
$k \geq 2$, in a proper k-binomial-coloring of a k-binomial graph, each color in $[k]$ is
assigned to exactly 2^{k-2} edges.

Fig. 8.6 A labeled proper
3-binomial-colored graph

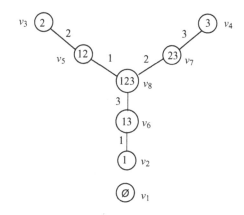

First, we show that there is a proper k-binomial-colored graph for every integer $k \geq 2$.

Theorem 8.5 ([21]). *For every integer $k \geq 2$, there exists a proper k-binomial-colored graph.*

Proof. We proceed by induction on k. We have already seen that there exists a proper 2-binomial-colored graph. Assume that there exists a proper k-binomial-colored graph H, where $V(H) = \{v_1, v_2, \ldots, v_{2^k}\}$, such that $\deg v_i \leq \deg v_j$ if $i \leq j$. If $\deg v_i = \deg v_j$ and $c'(v_i)$ precedes $c'(v_j)$ lexicographically, then $i < j$. So, for the graph G_4 of Fig. 8.5, the vertices are labeled as shown in Fig. 8.6.

Let H' be another copy of the graph H where the vertex v_i ($1 \leq i \leq 2^k$) in H is labeled v_i' in H'. Thus, the edge colorings of H and H' are identical. Let G be the graph obtained from H and H' by adding the 2^{k-1} edges $v_{2i-1}v_{2i}'$ for $1 \leq i \leq 2^{k-1}$ and assign the color $k+1$ to each of these 2^{k-1} edges. This is illustrated in Fig. 8.7 for $k = 3$.

The number of vertices of degree r in G for $0 \leq r \leq k$ is therefore the sum of the number $\binom{k}{r-1}$ of vertices of degree $r-1$ in H and the number $\binom{k}{r}$ of vertices of degree r in H'. Since $\binom{k}{r-1} + \binom{k}{r} = \binom{k+1}{r}$, it follows that G has $\binom{k+1}{r}$ vertices of degree r. Furthermore, the resulting $(k+1)$-edge coloring of G is a proper $(k+1)$-binomial-coloring of G and so G is a proper $(k+1)$-binomial-colored graph.

By the Principle of Mathematical Induction, there exists a proper k-binomial-colored graph for every integer $k \geq 2$. □

We mentioned that edge colorings of graphs were introduced by Tait [76] when he used proper 3-edge colorings (later called *Tait colorings*) of 3-regular bridgeless planar graphs to generate 4-region colorings of these graphs. In fact, the first theoretical paper on graph theory occurred in an 1891 article of Petersen [72] that also dealt with regular graphs. This suggested applying binomial colorings to regular graphs. Necessarily, these graphs can contain no isolated vertices. In this case, we therefore no longer restrict our attention to proper edge colorings.

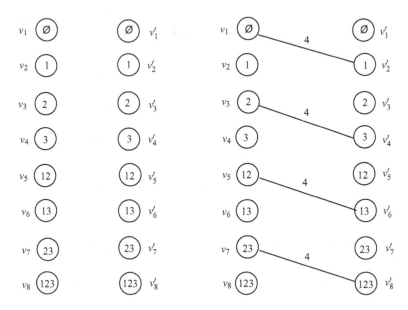

Fig. 8.7 Illustrating a step of the proof of Theorem 8.5

An unrestricted edge coloring $c : E(G) \rightarrow [k], k \geq 2$, of a graph G is a *k-binomial coloring* of G if the induced vertex coloring $c' : V(G) \rightarrow \mathscr{P}([k])$, where $c'(v)$ is the set of colors of the edges incident with v, satisfies the condition

$$\{c'(v) : v \in V(G)\} = \mathscr{P}^*([k]) = \mathscr{P}([k]) - \{\emptyset\}.$$

A graph G is an *unrestricted k-binomial-colored graph* (or simply a *k-binomial-colored graph*) in this case if G has an (unrestricted) k-binomial coloring. Here, we are interested in, for a fixed integer $k \geq 2$, the existence of an r-regular k-binomial graph of order n, where r and/or n is as small as possible. Necessarily, $r \geq k$ and $n \geq 2^k - 1$. These concepts were also introduced and studied in [21]. The following results were obtained.

Theorem 8.6 ([21]). *For each integer $k \geq 2$, there exists a k-regular k-binomial-colored graph of order 2^k.*

Proof. We show, in fact, that the k-regular k-cube $G = Q_k$ of order 2^k is a k-binomial-colored graph. The vertices of G can be labeled by the set of k-bit sequences. For $0 \leq \ell \leq k$, let V_ℓ be the set consisting of the $\binom{k}{\ell}$ vertices of G whose k-bit labels have exactly ℓ terms equal to 1. Thus, $V(G) = \bigcup_{\ell=0}^{k} V_\ell$. For $0 \leq \ell \leq k$, let

$$V_\ell = \left\{ v_{\ell,1}, v_{\ell,2}, \ldots, v_{\ell,\binom{k}{\ell}} \right\},$$

where the vertices of each set V_ℓ are listed in reverse lexicographical order.

For example, for $k = 3$, it follows that

$$V_0 = \{v_{0,1}\}, \quad V_1 = \{v_{1,1}, v_{1,2}, v_{1,3}\}, \quad V_2 = \{v_{2,1}, v_{2,2}, v_{2,3}\}, \quad V_3 = \{v_{3,1}\},$$

where

$$v_{0,1} = (0,0,0), \quad v_{1,1} = (1,0,0), \quad v_{1,2} = (0,1,0), \quad v_{1,3} = (0,0,1),$$

$$v_{2,1} = (1,1,0), \quad v_{2,2} = (1,0,1), \quad v_{2,3} = (0,1,1), \quad v_{3,1} = (1,1,1).$$

Since two vertices u and v are adjacent in G if and only if the labels of u and v differ in exactly one position, it follows that one of u and v belongs to V_i and the other belongs to V_{i+1} for some i ($0 \le i \le k-1$), say $u \in V_i$ and $v \in V_{i+1}$, and that every term having the value 1 for u also has the value 1 for v.

It remains to show that G has an unrestricted edge coloring $c : E(G) \to [k]$ such that $\{c'(v) : v \in V(G)\} = \mathscr{P}^*([k])$. First, define $c(v_{0,1}v_{1,i}) = i$ for $i = 1, 2, \ldots, \binom{k}{1} = k$. Then $c'(v_{0,1}) = [k]$. Next, define $c(e_i) = i$ for each edge e_i incident with $v_{1,i}$ and so $c'(v_{1,i}) = \{i\}$ for $1 \le i \le k$.

Assume, for a fixed integer j with $2 \le j \le k-1$ and for all integers i with $2 \le i \le j$, that all edges joining a vertex in V_{i-1} and a vertex in V_i have been assigned colors by the coloring c so that $c'(v_{i-1,t})$, $1 \le t \le \binom{k}{i-1}$, is the subset of $[k]$ in which $s \in c'(v_{i-1,t})$ if and only if the sth coordinate of $v_{i-1,t}$ is 1. Furthermore, this is the case for $c'(v_{j,t})$ as well, where $1 \le t \le \binom{k}{j}$, taking into consideration only those edges joining $v_{j,t}$ to the vertices in V_{j-1}.

Next, let $x \in V_j$ and $y \in V_{j+1}$ such that $xy \in E(G)$. The labels of x and y therefore differ in exactly one coordinate. Let $x = (x_1, x_2, \ldots, x_k)$ and $y = (y_1, y_2, \ldots, y_k)$, where then exactly j of the coordinates of x have the value 1, exactly $j + 1$ of the coordinates of y have the value 1 and there is a unique integer r with $1 \le r \le k$ such that $x_r = 0$ and $y_r = 1$. If p is the largest integer, $1 \le p \le k$, such that $p < r$ and $x_p = y_p = 1$, then define $c(xy) = p$. If $r = 1$ or $r \ge 2$ and $x_i = y_i = 0$ for $1 \le i \le r-1$, then p is the largest integer for which $x_p = y_p = 1$. The coloring c is illustrated in Fig. 8.8 for $k = 4$ where each k-bit $(x_1, x_2, \ldots, x_k)$ is denoted by $x_1 x_2 \ldots x_k$.

It remains to show that for each vertex in $V_j \cup V_{j+1}$ where $j \le k-1$, the induced color of the vertex consists of the subscripts of those terms having value 1. First, let $x \in V_j$. Then $x = (x_1, x_2, \ldots, x_k)$ and exactly j of the terms $x_1, x_2, \ldots, x_k$ are 1. Suppose that these j terms are $x_{n_1}, x_{n_2}, \ldots, x_{n_j}$ where $1 \le n_1 < n_2 < \cdots < n_j$. Then the set of colors of the edges joining x with the vertices in V_{j-1} is $\{n_1, n_2, \ldots, n_j\}$. Since the color of any edge joining x and a vertex $y = (y_1, y_2, \ldots, y_k) \in V_{j+1}$ is some integer n_t for which $x_{n_t} = y_{n_t} = 1$, it follows that $c'(x) = \{n_1, n_2, \ldots, n_j\}$. Next, let $y = (y_1, y_2, \ldots, y_k) \in V_{j+1}$ and exactly $j + 1$ of the terms $y_1, y_2, \ldots, y_k$ are 1. Suppose that these $j + 1$ terms are $y_{m_1}, y_{m_2}, \ldots, y_{m_{j+1}}$, where $1 \le m_1 < m_2 < \cdots < m_{j+1} \le k$. By the defining property of c, each edge joining y and a vertex in V_j

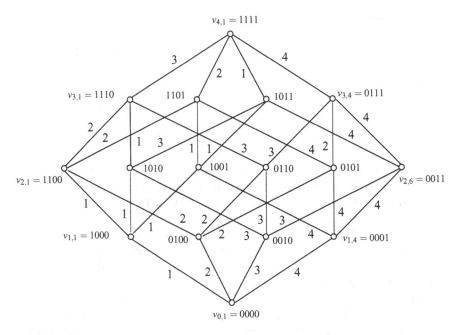

Fig. 8.8 Illustrating the coloring c in the proof of Theorem 8.6 for $k = 4$

is colored with some integer in $\{m_1, m_2, \ldots, m_{j+1}\}$. We now show that for each m_i with $1 \le i \le j + 1$, there is an edge joining y and a vertex $x \in V_j$ that is colored m_i by c. Let

$$x = (y_1, y_2, \ldots, y_{m_i}, \ldots, y_{m_{i+1}} - 1, \ldots, y_k).$$

Then x has exactly j terms having value 1 and so $x \in V_j$. The labels of x and y differ in exactly one position, namely the m_{i+1}th position, where $x_{m_{i+1}} = y_{m_{i+1}} - 1$. Since m_i is the largest integer such that $x_{m_i} = y_{m_i} = 1$, it follows that $c(xy) = m_i$. Hence, $c'(y) = \{m_1, m_2, \ldots, m_{j+1}\}$, taking into consideration only those edges joining y to the vertices in V_j. Therefore, c is a k-binomial coloring of G and so G is a k-binomial-colored graph. $\qquad\square$

With the aid of the proof of Theorem 8.6, the following can be proved.

Theorem 8.7 ([21]). *Let $k \ge 4$ be an integer.*

- *If k is even, then there exists a k-regular k-binomial-colored graph of order $2^k - 1$.*
- *If k is odd, then there exists a $(k + 1)$-regular k-binomial-colored graph of order $2^k - 1$.*

Every k-binomial graph we have seen possesses a k-binomial-coloring. Consequently, this suggests the following conjecture.

Conjecture 8.8 ([21]). *For each integer $k \ge 2$, every k-binomial graph is a k-binomial-colored graph.*

8.3   The Neighbor Strong Chromatic Index

In Sect. 8.1, we discussed proper edge colorings whose colors are elements of a set $[k]$ for an integer $k \geq 2$ and which generate vertex colorings in which the color of a vertex is the set of colors of its incident edges. If every two vertices are colored differently, then we saw that the edge coloring is a vertex-distinguishing strong edge coloring or, more simply, a strong edge coloring. On the other hand, if the edge coloring only requires every two adjacent vertices to be colored differently (and so the vertex coloring is also a proper coloring), then the edge coloring is called a *neighbor strong edge coloring*. Consequently, if c is a neighbor strong edge coloring of a connected graph G of order at least 3 and $S(v)$ is the set of colors of the edges incident with a vertex v of G, then $S(u) \neq S(v)$ for every two adjacent vertices u and v of G.

The minimum positive integer k for which G has a neighbor strong k-edge coloring is called the *neighbor strong chromatic index* of G, which is denoted by $\chi'_{ns}(G)$. These concepts were introduced by Zhang et al. [81] who referred to this concept as the *adjacent strong chromatic number* of G. Since (a) every neighbor strong edge coloring of a graph G is a proper edge coloring of G and (b) the neighbor strong chromatic index of G can never exceed its strong chromatic index, it follows that

$$\Delta(G) \leq \chi'(G) \leq \chi'_{ns}(G) \leq \chi'_s(G). \tag{8.1}$$

To illustrate the neighbor strong chromatic index, we determine $\chi'_{ns}(G)$ for the graph G shown in Fig. 8.1. Since $\chi'_s(G) = 5$ for this graph G, it follows by (8.1) that $\chi'_{ns}(G) \leq 5$. Furthermore, since $\chi'(G) = \Delta(G) = 3$, it follows by (8.1) that $\chi'_{ns}(G) \geq 3$. First, we show that $\chi'_{ns}(G) \neq 3$. If $\chi'_{ns}(G) = 3$, then there is a proper 3-edge coloring of G using the colors 1, 2, 3, say, that is neighbor-distinguishing. However, v and w are two adjacent vertices of degree 3 and so $S(v) = S(w) = \{1, 2, 3\}$, which is impossible. Since the proper 4-edge coloring of G shown in Fig. 8.9 is neighbor-distinguishing, it follows that $\chi'_{ns}(G) = 4$.

Fig. 8.9 A graph G with $\chi'_{ns}(G) = 4$

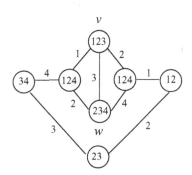

While showing that the neighbor strong chromatic index of the graph G in Fig. 8.9 is 4, we noted that since G contains two adjacent vertices of degree $3 = \Delta(G)$, it follows that $\chi'_{ns}(G) \geq 4$. This is a special case of the following.

Observation 8.9. *If a connected graph G of order 3 or more contains two adjacent vertices of degree $\Delta(G)$, then $\chi'_{ns}(G) \geq \Delta(G) + 1$.*

To illustrate the neighbor strong chromatic index further, we describe some results concerning the neighbor strong chromatic index of trees beginning with double stars. For integers $a, b \geq 2$, recall that the double star $S_{a,b}$ is the tree containing exactly two vertices (necessarily adjacent), called the *central vertices*, that are not end-vertices, one of degree a and the other of degree b.

Proposition 8.10. *For the double star $S_{a,b}$, where $a, b \geq 2$,*

$$\chi'_{ns}(S_{a,b}) = \begin{cases} \Delta(S_{a,b}) & \text{if } a \neq b \\ \Delta(S_{a,b}) + 1 & \text{if } a = b. \end{cases}$$

Proof. Let u and v be central vertices such that u is adjacent to the end-vertices u_1, $u_2, \ldots, u_{a-1}$ and v is adjacent to the end-vertices $v_1, v_2, \ldots, v_{b-1}$. We may assume that $a \leq b$ and so $\Delta(S_{a,b}) = b$. If $a < b$, then define the coloring $c : E(S_{a,b}) \to [b]$ of $S_{a,b}$ by

$$c(e) = \begin{cases} 1 & \text{if } e = uv \\ i + 1 & \text{if } e = uu_i, \, 1 \leq i \leq a - 1 \\ j + 1 & \text{if } e = vv_j, \, 1 \leq j \leq b - 1. \end{cases}$$

Hence, no two adjacent vertices have the same degree and so this proper b-edge coloring of $S_{a,b}$ is a neighbor strong edge coloring. Thus, $\chi'_{ns}(S_{a,b}) = \Delta(S_{a,b}) = b$ when $a < b$.

If $a = b$, then define the coloring $c : E(S_{a,b}) \to [b + 1]$ of $S_{a,b}$ by

$$c(e) = \begin{cases} 1 & \text{if } e = uv \\ i + 1 & \text{if } e = uu_i, \, 1 \leq i \leq a - 1 \\ j + 2 & \text{if } e = vv_j, \, 1 \leq j \leq b - 1. \end{cases}$$

Hence, the only two adjacent vertices of the same degree are u and v, and v is incident with an edge colored $b + 1$ while u is not. Thus, c is a neighbor strong $(b + 1)$-edge coloring. By Observation 8.9, $\chi'_{ns}(S_{a,b}) = \Delta(S_{a,b}) + 1 = b + 1$ when $a = b$. $\square$

Theorem 8.11 ([81]). *If T is a tree of order at least 3, then*

$$\Delta(T) \leq \chi'_{ns}(T) \leq \Delta(T) + 1.$$

Furthermore, $\chi'_{ns}(T) = \Delta(T)$ if and only if no two vertices of maximum degree are adjacent.

Proof. Let T be a tree of order $n \geq 3$. First, we show that $\chi'_{ns}(T) \leq \Delta(T) + 1$. We proceed by induction on n. The statement is obviously true for the path P_3 of order 3. Assume that the statement is true for all trees of order k where $k \geq 3$ and let T be a tree of order $k + 1$. Because the theorem is true for stars, we may assume that T is not a star. Thus, T contains an end-vertex v such that $\Delta(T) = \Delta(T - v)$. Let $T' = T - v$. By the induction hypothesis, there is a neighbor strong edge coloring c of T' using $\Delta(T) + 1$ colors. Let u be the vertex of T that is adjacent to v. Since $\deg_T u \leq \Delta(T) - 1$, there is a color $a \in [\Delta(T) + 1]$ that is not used by c to color any edge incident with u. Assigning a to the edge uv, we obtain a neighbor strong edge coloring of T using at most $\Delta(T) + 1$ colors and so $\chi'_{ns}(T) \leq \Delta(T) + 1$.

If T contains two adjacent vertices of maximum degree, then $\chi'_{ns}(T) \geq \Delta(T) + 1$ by Observation 8.9 and so $\chi'_{ns}(T) = \Delta(T) + 1$ in this case.

It remains to show that if no two vertices of maximum degree are adjacent, then $\chi'_{ns}(T) = \Delta(T)$. Again, we proceed by induction on n. First, observe that the statement is true for all trees of order n, where $3 \leq n \leq 6$, in which no two vertices of maximum degree are adjacent. Assume that the statement is true for all such trees of order k where $k \geq 6$ and let T be a tree of order $k + 1$ in which no two vertices of maximum degree are adjacent. By Proposition 8.10, we may also assume that T is neither a star nor a double star. Let v be an end-vertex of T on a longest path P in T where (v, u, w) is a subpath in P. Thus, $\deg u \geq 2$ and $\deg w \geq 2$. Let $T' = T - v$. Then no two vertices of degree $\Delta(T)$ are adjacent in T'; however, T' may contain two adjacent vertices of degree $\Delta(T')$. If T' contains two adjacent vertices of degree $\Delta(T')$, then $\Delta(T') = \Delta(T) - 1$ and u is the one vertex of T for which $\deg u = \Delta(T)$. In this case, we select the other end-vertex of P as u. Consequently, we may assume that $\Delta(T') = \Delta(T)$ and T' has no two adjacent vertices of degree $\Delta(T)$. By the induction hypothesis, there is a neighbor strong edge coloring c of T' using the $\Delta(T)$ colors in the set $[\Delta(T)] = \{1, 2, \ldots, \Delta(T)\}$. Next we define an edge coloring c_T of T from c as follows.

For $z \in V(T')$, let $S'(z)$ denote the set of colors (produced by the edge coloring c of T') of the edges incident with z in T' and let $S(z)$ denote the set of colors (produced by an edge coloring of T) of the edges incident with z in T. The vertex w is the unique neighbor of u for which $\deg w \geq 2$. Then $\deg w \leq \Delta(T)$ and so $|S'(w)| \leq \Delta(T)$. Also, $|S'(u)| \leq \Delta(T) - 1$. Since $uw \in E(T)$, at most one of u and w has degree $\Delta(T)$ in T. Suppose first that $S'(u) - S'(w) \neq \emptyset$ and $S'(w) - S'(u) \neq \emptyset$. Then there exist $a, b \in [\Delta(T)]$ such that $a \in S'(w) - S'(u)$ and $b \in S'(u) - S'(w)$. Since $|S'(u)| \leq \Delta(T) - 1$, there is a color $d \in [\Delta(T)]$ such that $d \notin S'(u)$. Assigning the color d to the edge uv results in a new coloring c_T of T such that $S(w) \neq S(u)$, $S(v) = \{d\} \neq S(u)$ and $S(x) = S'(x)$ for all $x \in V(T) - \{w, u, v\}$. Thus, $\chi'_{ns}(T) = \Delta(T)$. Consequently, we may assume that either $S'(w) \subset S'(u)$ or $S'(u) \subset S'(w)$. We now consider these two cases.

Case 1. $S'(w) \subset S'(u)$. Thus $\deg w = \deg_{T'} w < \deg_{T'} u \leq \Delta(T) - 1$. Then there exists $a \in [\Delta(T)]$ such that $a \notin S'(u)$. Assigning the color a to the edge uv results in a new coloring of T such that $S(w) \neq S(u)$, $S(v) = \{a\} \neq S(u)$ and

$S(x) = S'(x)$ for all $x \in V(T) - \{w, u, v\}$. Thus, this new coloring c_T is a neighbor strong edge coloring of T using $\Delta(T)$ colors and so $\chi'_{ns}(T) = \Delta(T)$.

Case 2. $S'(u) \subset S'(w)$. Then $\deg_{T'} u \leq \deg_{T'} w = \deg w$. We consider two subcases.

Subcase 2.1. $\deg_{T'} w \leq \Delta(T) - 1$. Hence, there exists $a \in [\Delta(T)]$ such that $a \notin S'(w)$. Assigning the color a to the edge uv and proceeding as in Case 1, we see that $\chi'_{ns}(T) = \Delta(T)$.

Subcase 2.2. $\deg_{T'} w = \Delta(T)$. Then $|S'(u)| = \deg_{T'} u \leq \Delta(T) - 2$. Let $a \in [\Delta(T)]$ such that $a \notin S'(u)$. Assign the color a to the edge uv. Since $|S(w)| = |S'(w)| = \Delta(T)$ and $|S(u)| \leq \Delta(T) - 1$, it follows that $S(w) \neq S(u)$. Proceeding as in Case 1, we obtain $\chi'_{ns}(T) = \Delta(T)$. $\qquad\qquad\square$

We saw in Corollary 1.7 that if the set of vertices of maximum degree in a graph G is an independent set, then G is a class one graph and so $\chi'(G) = \Delta(G)$. Thus, we have the following result.

Theorem 8.12. *If no two adjacent vertices of a connected graph G have the same degree, then $\chi'_{ns}(G) = \Delta(G)$.*

Chapter 9
Sum-Defined Chromatic Indices

In this chapter, proper edge colorings of graphs are considered in which the colors are either positive integers or elements of $\mathbb{Z}_k$ for some positive integer k. In the first instance, these edge colorings give rise to either irregular vertex colorings or proper vertex colorings; while in the second instance, the edge colorings give rise to proper vertex colorings. In both instances, however, the color of a vertex is the sum of the colors of its incident edges. We begin with edge colorings in which the colors are positive integers.

9.1 The Irregular-Sum Chromatic Index

For a connected graph G of order at least 3, a proper edge coloring $c : E(G) \to [k]$ for some integer $k \geq 2$ is called an *irregular-sum edge coloring* of G if the induced vertex coloring $c' : V(G) \to \mathbb{N}$ defined by

$$c'(v) = \sum_{e \in E_v} c(e)$$

is irregular (vertex-distinguishing). The minimum positive integer k for which G has such an irregular-sum edge coloring is called the *irregular-sum chromatic index* of G and is denoted by $\chi'_{is}(G)$. Thus, if G is a connected graph of size at least 2, then

$$\Delta(G) \leq \chi'(G) \leq \chi'_{is}(G).$$

These concepts were studied by Mahéo and Saclé in [70]. First, we present a useful observation.

Observation 9.1. *If G is a connected graph of order at least 3 such that G contains two adjacent vertices of maximum degree, then $\chi'_{is}(G) \geq \Delta(G) + 1$.*

© Ping Zhang 2015
P. Zhang, *Color-Induced Graph Colorings*, SpringerBriefs in Mathematics,
DOI 10.1007/978-3-319-20394-2_9

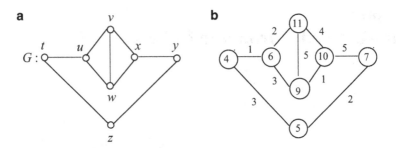

Fig. 9.1 A graph G with $\chi'_{is}(G) = 5$

As an illustration of the irregular-sum chromatic index, we determine $\chi'_{is}(G)$ for the graph G in Fig. 9.1a. Since G contains adjacent vertices having maximum degree $\Delta(G) = 3$, it follows by Observation 9.1 that $\chi'_{is}(G) \geq 4$. We show that $\chi'_{is}(G) \neq 4$. Suppose that there exists an irregular-sum 4-edge coloring of G. Thus every vertex of G must have one of the induced colors $3, 4, \ldots, 9$. Since there are 7 vertex colors in all and the order of G is 7, each of these colors is the color of exactly one vertex of G. Furthermore, the colors of the three vertices of degree 2 must be 3, 4 and 5 and the colors of the four vertices of degree 3 must be 6, 7, 8 and 9. Since the only way to express 6, 7, 8 and 9 as the sum of three distinct positive integers among $1, 2, 3, 4$ is

$$6 = 1 + 2 + 3, \ 7 = 1 + 2 + 4, \ 8 = 1 + 3 + 4 \text{ and } 9 = 2 + 3 + 4,$$

it follows that for each color $a \in \{1, 2, 3, 4\}$, three vertices among u, v, w, x are incident with an edge colored a. Thus, there must be two edges colored with each of $1, 2, 3, 4$ among the edges incident with any of u, v, w, x. Since there are only 7 such edges, this is impossible. Thus $\chi'_{is}(G) \geq 5$. The irregular-sum 5-edge coloring of G in Fig. 9.1b shows that $\chi'_{is}(G) = 5$.

The irregular-sum chromatic indices of complete graphs, paths and cycles were determined by Mahéo and Saclé [70].

Theorem 9.2 ([70]). *If $n \geq 3$, then*

$$\chi'_{is}(K_n) = \begin{cases} n & \text{if } n \text{ is odd} \\ n + 1 & \text{if } n \text{ is even.} \end{cases}$$

Theorem 9.3 ([70]). *If $n \geq 3$, then*

$$\chi'_{is}(P_n) = \begin{cases} \frac{n+2}{2} & \text{if } n \text{ is even} \\ \frac{n+3}{2} & \text{if } n \equiv 1 \pmod 4 \\ \frac{n+1}{2} & \text{if } n \equiv 3 \pmod 4 \text{ and } n \neq 7 \\ 5 & \text{if } n = 7. \end{cases}$$

Theorem 9.4 ([70]). *If $n \geq 3$, then*

$$\chi'_{is}(C_n) = \begin{cases} \frac{n+4}{2} & \text{if } n \text{ is even} \\ \frac{n+5}{2} & \text{if } n \equiv 1 \pmod 4 \\ \frac{n+3}{2} & \text{if } n \equiv 3 \pmod 4. \end{cases}$$

The irregular-sum chromatic indices of the complete bipartite graphs $K_{r,s}$, where $2 \leq r \leq s$, were also studied in [70] and the following result was obtained.

Theorem 9.5 ([70]). *Let r and s be integers with $2 \leq r \leq s$. If (i) $s = r$ or (ii) $s = r + 1$ and $r \leq 7$, then $\chi'_{is}(K_{r,s}) = r + 2$.*

The values of $\chi'_{is}(K_{r,s})$ were determined in [70] for several other pairs r, s of integers with $2 \leq r \leq s$. In fact, there is a conjecture in this connection.

Conjecture 9.6 ([70]). Let r and s be integers with $2 \leq r \leq s$.

(1) If $2 \leq r \leq s - 2$, then $\chi'_{is}(K_{r,s}) = s + 1$.
(2) If $r = s - 1 \geq 8$, then $\chi'_{is}(K_{r,s}) = s + 2$.

A lower bound for the irregular-sum chromatic index of a connected graph G of order at least 3 was established in terms of the maximum degree, minimum degree and the order of G.

Theorem 9.7 ([70]). *If G is a connected graph of order $n \geq 3$ having maximum degree Δ and minimum degree δ, then*

$$\chi'_{is}(G) \geq \left\lceil \frac{n-1}{\Delta} + \frac{\Delta - 1}{2} + \frac{\delta(\delta + 1)}{2\Delta} \right\rceil. \tag{9.1}$$

Proof. Let $\chi'_{is}(G) = k$ and let c be an irregular-sum k-edge coloring of G. Furthermore, let

$$\alpha = \min\{c'(v) : v \in V(G)\} \text{ and } \beta = \max\{c'(v) : v \in V(G)\}.$$

Thus, $\beta - \alpha \geq n - 1$. Since

$$\alpha \geq 1 + 2 + \cdots + \delta = \frac{\delta(\delta + 1)}{2} \text{ and}$$

$$\beta \leq (k - \Delta + 1) + (k - \Delta + 2) + \cdots + k = \frac{\Delta(2k - \Delta + 1)}{2},$$

it follows that

$$n - 1 \leq \beta - \alpha \leq \frac{\Delta(2k - \Delta + 1)}{2} - \frac{\delta(\delta + 1)}{2},$$

resulting in the inequality in (9.1). $\qquad\square$

By Theorem 9.5, $\chi'_{is}(K_{r,r+1}) = r + 2$ for $2 \leq r \leq 7$ and so the lower bound in (9.1) is attained for these graphs. In the case of regular graphs, this lower bound can be improved slightly.

Theorem 9.8. *If G is an r-regular connected graph of order $n \geq 3$, then*

$$\chi'_{is}(G) \geq \left\lceil r + \frac{n-1}{r} + \frac{2\epsilon}{nr} \right\rceil, \tag{9.2}$$

where $\epsilon = 1$ if $n[r(r+1) + n - 1]/2$ is an odd integer and $\epsilon = 0$ otherwise.

It is known that the irregular-sum chromatic index of the Petersen graph is 7. Consequently, the lower bound in (9.2) is attained in this case.

9.2 The Proper-Sum Chromatic Index

Let G be a connected graph of order at least 3. A proper edge coloring $c : E(G) \rightarrow [k]$ for some integer $k \geq 2$ is called a *proper-sum edge coloring* of G if the induced vertex coloring $c' : V(G) \rightarrow \mathbb{N}$ defined by

$$c'(v) = \sum_{e \in E_v} c(e)$$

is proper as well. The minimum positive integer k for which G has such a proper-sum edge coloring is called the *proper-sum chromatic index* of G and is denoted by $\chi'_{ps}(G)$. These concepts were studied by Flandrin et al. [42] using the terminology *neighbor sum distinguishing edge coloring* and *neighbor sum distinguishing index*. For every connected graph G of order at least 3,

$$\Delta(G) \leq \chi'(G) \leq \chi'_{ps}(G) \leq \chi'_{is}(G). \tag{9.3}$$

Since the result stated in Observation 9.1 holds in this case as well, we have the following.

Observation 9.9. *If G is a connected graph of order at least 3 such that G contains two adjacent vertices of maximum degree, then $\chi'_{ps}(G) \geq \Delta(G) + 1$.*

To illustrate the proper-sum chromatic index, we determine $\chi'_{ps}(G)$ for the graph G in Fig. 9.2. Since $\Delta(G) = 3$, it follows by Observation 9.9 that $\chi'_{ps}(G) \geq 4$. Since the proper-sum 4-edge coloring of G shown in Fig. 9.2 is neighbor-distinguishing, it follows that $\chi'_{ps}(G) = 4$.

We next consider the graph $C_4 = (v_1, v_2, v_3, v_4, v_1)$. Since $\Delta(C_4) = 2$ and C_4 contains adjacent vertices of degree 2, it follows by Observation 9.9 that $\chi'_{ps}(C_4) \geq 3$. If $\chi'_{ps}(C_4) = 3$, then there exists a proper-sum edge coloring c with the colors $1, 2, 3$. However then, two nonadjacent edges, say $v_1 v_2$ and $v_3 v_4$, are colored

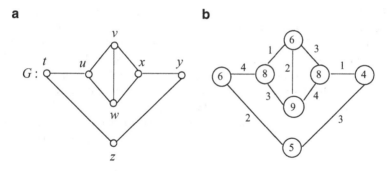

Fig. 9.2 A graph G with $\chi'_{ps}(G) = 4$

the same, implying that $c'(v_1) = c'(v_4)$ and $c'(v_2) = c'(v_3)$, which is impossible. The proper-sum 4-edge coloring of C_4 that assigns the color i to the edge $v_i v_{i+1}$ for $i = 1, 2, 3, 4$ (where $v_5 = v_1$) shows that $\chi'_{ps}(C_4) = 4$.

Also, suppose that $C_5 = (v_1, v_2, v_3, v_4, v_5, v_1)$. Again, it follows by Observation 9.9 that $\chi'_{ps}(C_5) \geq 3$. In fact, $\chi'_{ps}(C_5) \geq 5$, for if there exists a proper-sum 4-edge coloring c of C_5 using the colors $1, 2, 3, 4$, then two nonadjacent edges, say $v_1 v_2$ and $v_3 v_4$, are colored the same. However then $c'(v_2) = c'(v_3)$, a contradiction. Since the edge coloring of C_5 that assigns the edge $v_i v_{i+1}$ the color i for $1 \leq i \leq 5$ (where $v_6 = v_1$) is a proper-sum 5-edge coloring, it follows that $\chi'_{ps}(C_5) = 5$. As the following conjecture suggests, the graph C_5 may be the sole exception to the belief that $\Delta(G) \leq \chi'_{ps}(G) \leq \Delta(G) + 2$ for every connected graph G of order at least 3.

Conjecture 9.10 ([42]). If G is a connected graph of order at least 3 and $G \neq C_5$, then $\chi'_{ps}(G) \leq \Delta(G) + 2$.

Since $\chi'_{is}(K_n) = \chi'_{ps}(K_n)$ for each integer $n \geq 3$, the following is a consequence of Theorem 9.2.

Theorem 9.11 ([42]). *For each integer $n \geq 3$,*

$$\chi'_{ps}(K_n) = \begin{cases} n & \text{if } n \text{ is odd} \\ n + 1 & \text{if } n \text{ is even.} \end{cases}$$

The proper-sum chromatic indices of complete bipartite graphs, paths and cycles were determined in [42]. As an illustration, we verify the formula for $\chi'_{ps}(K_{r,s})$.

Theorem 9.12 ([42]). *For integers r and s with $1 \leq r \leq s$ and $s \geq 2$,*

$$\chi'_{ps}(K_{r,s}) = \begin{cases} s + 2 & \text{if } r = s \\ s & \text{if } r < s. \end{cases}$$

Proof. First, assume that $r = s \geq 2$. Since $\chi'_{is}(K_{r,r}) = r + 2$ for each $r \geq 2$ by Theorem 9.5, it follows by (9.3) that $\chi'_{ps}(K_{r,r}) \leq r + 2$. By Observation 9.9, $\chi'_{ps}(K_{r,r}) \geq r + 1$. We show that $\chi'_{ps}(K_{r,r}) \neq r + 1$. Assume, to the contrary, that $K_{r,r}$ has a proper-sum $(r + 1)$-edge coloring c using the colors $1, 2, \ldots, r + 1$. Since $(r + 1) \nmid r^2$, there is a color in $[r + 1]$, say the color 1, such that the set of all edges colored 1 by c is not a perfect matching in $K_{r,r}$. Hence there are two adjacent vertices u and v such that no edge incident with u or v is colored 1. However then, $c'(u) = c'(v)$, which is impossible. Therefore, $\chi'_{ps}(K_{r,r}) = r + 2$. Next, assume that $r < s$. Since every proper s-edge coloring of $K_{r,s}$ is a proper-sum s-edge coloring, $\chi'_{ps}(K_{r,s}) = s$. $\qquad\square$

Theorem 9.13 ([42]). *For each integer $n \geq 3$,*

$$\chi'_{ps}(P_n) = \begin{cases} 2 & \text{if } n = 3 \\ 3 & \text{if } n \geq 4 \end{cases}$$

$$\chi'_{ps}(C_n) = \begin{cases} 3 & \text{if } n \equiv 0 \pmod 3 \\ 4 & \text{if } n \not\equiv 0 \pmod 3 \text{ and } n \neq 5 \\ 5 & \text{if } n = 5. \end{cases}$$

An upper bound for $\chi'_{ps}(G)$ in terms of $\Delta(G)$ was established in [42].

Theorem 9.14 ([42]). *If G is a connected graph of size at least 2, then*

$$\chi'_{ps}(G) \leq \left\lceil \frac{7\Delta(G) - 4}{2} \right\rceil.$$

Proof. We employ strong induction on the size $m \geq 2$ of a connected graph G to show that G has a proper-sum k-edge coloring, where $k = \lceil (7\Delta(G) - 4)/2 \rceil$. It is obvious that the statement is true for $m = 2, 3$. Assume that this statement holds for all connected graphs of size ℓ for each integer ℓ with $2 \leq \ell < m$ where $m \geq 4$ and let G be a connected graph of size m having maximum degree $\Delta = \Delta(G) \geq 2$. By Theorem 9.11, we may assume that G is not a complete graph and so G contains two vertices u and w with $d(u, w) = 2$. Thus, there exists a vertex v such that $uv, vw \in E(G)$ and $uw \notin E(G)$.

Let $H = G - \{uv, vw\}$. If H is connected, then we also denote H by H_1. If H is disconnected and has components K_1 and K_2, then we denote the union of all such components by H_0. Any components of H having size 2 or more are denoted by $H_1, H_2, \ldots, H_t$, where necessarily $t \leq 3$. By the induction hypothesis, each component H_i $(1 \leq i \leq t)$ has a proper-sum k-edge coloring c_i. We next construct a proper-sum k-edge coloring $c : E(G) \to [k]$ of G as follows. First, if H contains the component K_2, then define $c(e) = 1$ for each edge e in H_0. If H contains components H_i, $1 \leq i \leq t$, then define $c(e) = c_i(e)$ for each edge e in H_i. It now remains to show that there are colors $\alpha, \beta \in [k]$ such that $c(uv) = \alpha$, $c(vw) = \beta$ and the resulting k-edge coloring c is a proper-sum edge coloring.

Suppose first that all of $N_H(u)$, $N_H(v)$ and $N_H(w)$ are nonempty. Let

$$N_H(u) = \{u_1, u_2, \ldots, u_p\}, \ N_H(v) = \{v_1, v_2, \ldots, v_q\} \text{ and } N_H(w) = \{w_1, w_2, \ldots, w_r\},$$

where $1 \le p, r \le \Delta - 1$ and $1 \le q \le \Delta - 2$. Let U be the set consisting of the colors of the edges incident with u in H, let V be the set consisting of the colors of the edges incident with v in H and let W be the set consisting of the colors of the edges incident with w in H. Thus,

$$|U| = p \le \Delta - 1, \ |V| = q \le \Delta - 2 \text{ and } |W| = r \le \Delta - 1.$$

Since c is required to be a proper edge coloring of G, we must have

$$\alpha \in [k] - (\{\beta\} \cup U \cup V) \text{ and } \beta \in [k] - (\{\alpha\} \cup V \cup W).$$

Next, we consider the number of colors in $[k]$ that are available for α. For each vertex x of H, let $\sigma_H(x)$ denote the sum of the colors of the edges incident with x in H. The color α cannot be (1) any of the $q \ (\le \Delta - 2)$ colors of the edges vv_i for $1 \le i \le q$, (2) any of the $p \ (\le \Delta - 1)$ colors of the edges uu_i for $1 \le i \le p$ and (3) the color β. Furthermore, α must satisfy the following condition:

$$c'(u) = \sigma_H(u) + \alpha \ne c'(u_i) = \sigma_H(u_i) \text{ for } 1 \le i \le p.$$

That is, there are at most $(\Delta - 2) + 2(\Delta - 1) + 1 = 3(\Delta - 1)$ colors that cannot be chosen as α and so there are at least

$$k - 3(\Delta - 1) = \left\lceil \frac{7\Delta - 4}{2} \right\rceil - 3(\Delta - 1) = \left\lceil \frac{\Delta + 2}{2} \right\rceil = s$$

choices for α. Similarly, there are at least s choices for β.

Let X be the set of colors in $[k]$ that are available for α and let Y the set of colors that are available for β. As we observed, $|X| \ge s$ and $|Y| \ge s$. We show that there exists $(\alpha, \beta) \in X \times Y$ such that $\alpha \ne \beta$ and $\sigma_H(v) + \alpha + \beta \ne \sigma_H(v_i)$ for each i with $1 \le i \le q \le \Delta - 2$. To show that this is always possible, it is sufficient to observe that because $|X| \ge s$ and $|Y| \ge s$, there are at least $2s - 3$ pairs $(x_i, y_i) \in X \times Y$ (where $1 \le i \le 2s - 3$) such that $x_i \ne y_i$ and all of the sums $x_i + y_i$ are distinct. For example, let x be the minimum element of $X \cup Y$, say $x \in X$, and let y be the maximum element of Y. Then there are $2s - 3$ pairs in

$$[\{x\} \times (Y - \{x\})] \cup [(X - \{x, y\}) \times \{y\}]$$

having the desired property. Since $2s - 3 \ge (\Delta - 2) + 1$, there is at least one pair $(x_i, y_i) \in X \times Y$ that can be chosen for (α, β).

If some of $N_H(u)$, $N_H(v)$ and $N_H(w)$ are empty, the proof is similar. $\square$

This topic has also been discussed in many research papers (see [28, 79, 80], for example).

9.3   The Twin Chromatic Index

Let G be a connected graph of order at least 3. A proper edge coloring $c : E(G) \to \mathbb{Z}_k$ for some integer $k \geq 2$ is called a *twin edge coloring* if the induced vertex coloring $c' : V(G) \to \mathbb{Z}_k$ defined by

$$c'(v) = \sum_{e \in E_v} c(e)$$

is proper as well. The minimum integer $k \geq 2$ for which G has a twin k-edge coloring is called the *twin chromatic index* of G and is denoted by $\chi_t'(G)$. This concept was introduced by Chartrand and studied in [7–9]. Since a twin edge coloring is not only a proper edge coloring of G but induces a proper vertex coloring of G, it follows that

$$\chi_t'(G) \geq \max\{\chi(G), \chi'(G)\} \geq \Delta(G). \tag{9.4}$$

Since $\max\{\chi(G), \chi'(G)\} = \chi'(G)$ except when G is a complete graph of even order, we have $\chi_t'(G) \geq \chi'(G)$ except possibly when G is a complete graph of even order. As an illustration, Fig. 9.3 shows a twin edge 4-coloring c of the Petersen graph P, where the color $c'(v)$ is placed inside each vertex v of P and so $\chi_t'(P) \leq 4$. Since P is a 3-regular graph and P is not 1-factorable, $\chi'(P) = \Delta(P) + 1 = 4$. It then follows by (9.4) that $\chi_t'(P) = \chi'(P) = 4$. This example also illustrates the following useful observation, which we previously encountered in Observations 9.1 and 9.9 with regard to other sum-defined proper edge colorings.

Fig. 9.3   A twin edge 4-coloring of the Petersen graph

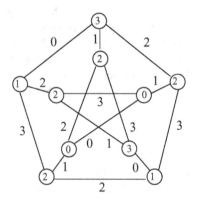

Observation 9.15. *If a connected graph G contains two adjacent vertices of degree $\Delta(G)$, then $\chi'_t(G) \geq 1 + \Delta(G)$. In particular, if G is a connected r-regular graph for some integer $r \geq 2$, then $\chi'_t(G) \geq 1 + r$.*

It is immediate that $\chi'_t(G)$ exists for every connected graph of order at least 3. For every connected graph G for which the twin chromatic index has been determined, we have seen that $\chi'_t(G) = \Delta(G) + i$ for some $i \in \{0, 1, 2, 3\}$. This leads us to the following problem:

Is $\chi'_t(G) \leq \Delta(G) + 3$ for every connected graph G of order at least 3?

We have seen that

$$\chi'_t(C_5) = 5 = \Delta(C_5) + 3.$$

For each connected graph that is not the 5-cycle, the following conjecture was made, which is reminiscent of Conjecture 9.10.

Conjecture 9.16 ([7]). If G is a connected graph of order at least 3 that is not a 5-cycle, then

$$\chi'_t(G) \leq \Delta(G) + 2.$$

Conjecture 9.16 was verified in [7, 9] for several well-known classes of graphs, including paths, cycles, complete graphs and complete bipartite graphs, which we state next.

Theorem 9.17 ([7, 9]). *If n, r, s are integers with $n \geq 3$, $1 \leq r \leq s$ and $s \geq 2$, then*

$$\chi'_t(P_n) = 3$$

$$\chi'_t(C_n) = \begin{cases} 3 & \text{if } n \equiv 0 \pmod 3 \\ 4 & \text{if } n \not\equiv 0 \pmod 3 \text{ and } n \neq 5 \\ 5 & \text{if } n = 5 \end{cases}$$

$$\chi'_t(K_n) = \begin{cases} n & \text{if } n \text{ is odd} \\ n+1 & \text{if } n \text{ is even} \end{cases}$$

$$\chi'_t(K_{r,s}) = \begin{cases} s & \text{if } s \geq r+2 \text{ and } r \geq 2 \\ s+1 & \text{if either } r = 1 \text{ and } s \not\equiv 1 \pmod 4 \\ & \quad \text{or } s = r+1 \geq 3 \\ s+2 & \text{if either } r = 1 \text{ and } s \equiv 1 \pmod 4 \\ & \quad \text{or } s = r \geq 2. \end{cases}$$

To illustrate proofs of the results stated in Theorem 9.17, we verify some values of $\chi_t'(K_n)$. First, assume that $n \geq 3$ is odd. By Observation 9.15, $\chi_t'(K_n) \geq 1 + \Delta(K_n) = n$. To show that $\chi_t'(K_n) \leq n$, let $V(K_n) = \{v_0, v_1, \ldots, v_{n-1}\}$ and arrange the vertices $v_0, v_1, \ldots, v_{n-1}$ consecutively in a regular n-gon and join every two vertices by a straight line segment, producing K_n. For each i ($0 \leq i \leq n-1$), assign to $v_{i-1}v_{i+1}$ and those edges parallel to $v_{i-1}v_{i+1}$ the color i. Then v_i has the color $\binom{n}{2} - i$, resulting in a proper vertex coloring of K_n. Thus $\chi_t'(K_n) = n$ when n is odd.

For even integers $n \geq 4$, we only consider the case when $n \equiv 0 \pmod 4$ since the proof for the case when $n \equiv 2 \pmod 4$ is lengthy.

Theorem 9.18. *If $n \geq 4$ is an integer with $n \equiv 0$ (mod 4), then*

$$\chi_t'(K_n) = n + 1.$$

Proof. First, we show that $\chi_t'(K_n) \geq n + 1$ for each even integer $n \geq 4$. Since $\chi_t'(K_n) \geq 1 + \Delta(K_n) = n$ by Observation 9.15, it remains to show that $\chi_t'(K_n) \neq n$. Assume, to the contrary, that $\chi_t'(K_n) = n$. Then there is a proper edge coloring of K_n using the colors in $\mathbb{Z}_n$ that results in a proper vertex coloring of K_n. Since every vertex of K_n has degree $n - 1$, the edges incident with each vertex of K_n are colored with an $(n - 1)$-element subset of $\mathbb{Z}_n$. For example, if v is a vertex of K_n, then there is exactly one element $a \in \mathbb{Z}_n$ that is not used in coloring the edges incident with v. Consequently, at most $\frac{n}{2} - 1$ edges of K_n are colored a, implying that there exists some other vertex u of K_n none of whose incident edges are colored a. However then, $c'(u) = c'(v) = \binom{n}{2} - a$, which is impossible since u and v are adjacent in K_n. Thus $\chi_t'(K_n) \geq n + 1$.

Next, we show that K_n has a twin edge $(n + 1)$-coloring when $n \equiv 0 \pmod 4$ and $n \geq 4$. Let $V(K_n) = \{v_0, v_1, \ldots, v_{n-1}\}$ and arrange the vertices $v_0, v_1, \ldots, v_{n-1}$ consecutively in a regular n-gon and join every two vertices by a straight line segment, thereby producing K_n.

Since $n \equiv 0 \pmod 4$ and $n \geq 4$, it follows that $n = 4k$ for some positive integer k. First, let $M_0, M_1, \ldots, M_{2k-1}$ be $2k$ pairwise edge-disjoint matchings of size $2k - 1$ in K_{4k} where each matching M_i ($0 \leq i \leq 2k - 1$) consists of those $2k - 1$ edges perpendicular to $v_i v_{2k+i}$. Then $H = K_{4k} - \left(\bigcup_{i=0}^{2k-1} M_i\right)$ is therefore a $(2k)$-regular graph. The graph H has a 1-factorization $\{F_1, F_2, \ldots, F_{2k}\}$ where F_i ($1 \leq i \leq 2k$) consists of the edge $v_i v_{i+1}$ and those edges parallel to $v_i v_{i+1}$. For $n = 8$ and $k = 2$, this is illustrated in Fig. 9.4.

Let

$$X_1 = \{v_0 v_{2k-1}, v_1 v_{2k-2}, \ldots, v_{k-1} v_k\} \text{ and } X_1' = \{v_{2k} v_{4k-1}, v_{2k+1} v_{4k-2}, \ldots, v_{3k-1} v_{3k}\}.$$

Thus $|X_1| = |X_1'| = k$ and $E(F_{k-1}) = X_1 \cup X_1'$. In particular, if $k = 2$, then $E(F_1) = X_1 \cup X_1'$ where $X_1 = \{v_0 v_3, v_1 v_2\}$ and $X_1' = \{v_4 v_7, v_5 v_6\}$. Define a coloring $c : E(K_{4k}) \to \mathbb{Z}_{4k+1}$ as follows. If $k = 2$, let

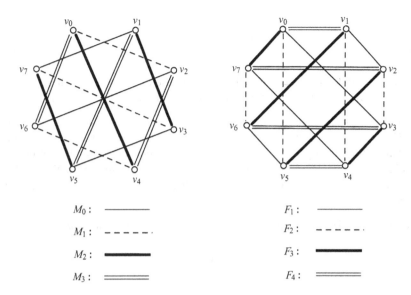

Fig. 9.4 Illustrating M_0, M_1, M_2, M_3 and F_1, F_2, F_3, F_4 for K_8

$$c(e) = \begin{cases} 0 & \text{if } e \in X_1' \\ i-1 & \text{if } e \in E(F_i) \text{ where } 2 \leq i \leq 2k \\ 2k & \text{if } e \in X_1 \\ 2k+j+1 & \text{if } e \in M_j \text{ where } 0 \leq j \leq 2k-1. \end{cases}$$

If $k \geq 3$, let

$$c(e) = \begin{cases} 0 & \text{if } e \in X_1' \\ i & \text{if } e \in E(F_i) \text{ where } 1 \leq i \leq k-2 \\ i-1 & \text{if } e \in E(F_i) \text{ where } k \leq i \leq 2k \\ 2k & \text{if } e \in X_1 \\ 2k+j+1 & \text{if } e \in M_j \text{ where } 0 \leq j \leq 2k-1. \end{cases} \tag{9.5}$$

Then c is a proper edge coloring. For $0 \leq i \leq 2k-1$,

$$c'(v_i) = \left[\binom{4k+1}{2} - 2k \right] - (2k+i+1) + 2k = -(2k+i+1) \text{ in } \mathbb{Z}_{4k+1};$$

while for $2k \leq i \leq 4k-1$,

$$c'(v_i) = \left[\binom{4k+1}{2} - 2k \right] - (i+1) + 0 = -(2k+i+1) \text{ in } \mathbb{Z}_{4k+1}.$$

Thus,

$$(c'(v_0), c'(v_1), \ldots, c'(v_{4k-1})) = (2k, 2k - 1, \ldots, 1, 0, 4k, 4k - 1, \ldots, 2k + 2).$$

That is, each color in $\mathbb{Z}_{4k+1}$ (except $2k + 1$) is used exactly once. Therefore, c' : $V(K_{4k}) \to \mathbb{Z}_{4k+1}$ is a proper vertex coloring of G and so $\chi'_t(K_n) = n + 1$.   □

Even though the results stated in Theorem 9.17 only reinforce the likelihood that Conjecture 9.16 is true, it has never even been verified that there is a positive constant C for which $\chi'_t(G) \leq \Delta(G) + C$ for every connected graph G of order at least 3. On the other hand, an upper bound for $\chi'_t(G)$ that is a linear function of $\Delta(G)$ has been established. This bound and a possible method of proof were suggested by Noga Alon.

Theorem 9.19 ([58]). *For every connected graph G of order at least 3,*

$$\chi'_t(G) \leq 4\Delta(G) - 3.$$

9.4 Trees

While little progress has been made in verifying Conjecture 9.16, this conjecture has been verified for trees with small maximum degree.

Theorem 9.20. *If T is a tree of order at least 3 with $\Delta(T) \leq 6$, then*

$$\chi'_t(T) \leq \Delta(T) + 2.$$

Proof. Let T be a tree of order at least 3 with $\Delta(T) \leq 6$. By Theorem 9.17, the result is true for paths. Thus we may assume that $\Delta(T) \geq 3$. We verify this theorem only when $\Delta(T) = 3$. The proofs when $4 \leq \Delta(T) \leq 6$, although more complex, employ similar techniques. Suppose that T is a tree with $\Delta(T) = 3$. If $T = K_{1,3}$, then the three edges of T can be colored 0, 1, 2 from $\mathbb{Z}_4$ or $\mathbb{Z}_5$, resulting in an induced color of 3 for the vertex of degree 3 in T. Since the induced colors of the leaves are 0, 1, 2, this coloring is a twin edge coloring of T and so $\chi'_t(T) \leq 5$. Since $\chi'_t(K_{1,3}) = 4$, we may assume that $T \neq K_{1,3}$.

Let $v \in V(T)$ such that $\deg_T v = 3$ and let $N_T(v) = \{v_1, v_2, v_3\}$. Since $T \neq K_{1,3}$, we may assume that v_1 is not a leaf. For a nonnegative integer i, let

$$A_i = \{u \in V(T) : d(v, u) = i\}.$$

Hence $A_0 = \{v\}$, $A_1 = \{v_1, v_2, v_3\}$ and $A_i \neq \emptyset$ for $i = 2$ and possibly for $i \geq 3$. Furthermore, if $A_i \neq \emptyset$ for some $i \geq 3$, then $A_{i-1} \neq \emptyset$ as well. Indeed, for each vertex $x \in A_i$, $i \geq 1$, there exists exactly one vertex $y \in A_{i-1}$ such that $xy \in E(T)$.

We now construct a twin 5-edge coloring c of T. First, define $c(vv_i) = i$ for $i = 1, 2, 3$. Thus, $c'(v) = 1$. Furthermore, if v_i is a leaf ($i = 2, 3$) of T, then $c'(v) \neq c(vv_i) = c'(v_i)$. Assume for a positive integer k that (1) a proper 5-edge coloring c has been defined for all edges wz of T for each edge of the subtree T_k of T whose vertex set is $A_0 \cup A_1 \cup \cdots \cup A_k$ such that $c(wz) \neq 0$ for each edge wz of T_k (with the only possible exception if wz is a pendant edge), (2) the induced vertex 5-coloring c' for all vertices in $A_0 \cup A_1 \cup \cdots \cup A_{k-1}$ is a proper vertex coloring and (3) if $wz \in E(T)$ such that $w \in A_{k-1}$, $z \in A_k$ and z is a leaf, then $c'(w) \neq c(wz) = c'(z)$. We now consider T_{k+1} and describe how $c(xw)$ can be defined for each edge xw where $x \in A_k$ and $w \in A_{k+1}$. Let y be the unique vertex in A_{k-1} that is adjacent to x. Then $c'(y) \in \{0, 1, 2, 3, 4\}$ and $c(xy)$ has been defined. We consider five cases, depending on the value of $c'(y)$. Case i ($i = 0, 1, 2, 3, 4$) will describe the situation where $c'(y) = i$. We begin with Case 0.

Case 0. $c'(y) = 0$. First, suppose that $\deg_T x = 2$ and x is adjacent to $w \in A_{k+1}$. If $c(xy) \in \{1, 4\}$, then define $c(xw) = 2$; while if $c(xy) \in \{2, 3\}$, then define $c(xw) = 1$. Then $c'(x) \neq c'(y)$. Furthermore, $c'(x) \neq c(xw)$. Next, suppose that $\deg_T x = 3$ and x is adjacent to the two vertices w and z in A_{k+1}. If $c(xy) \in \{1, 4\}$, then define $\{c(xw), c(xz)\} = \{2, 3\}$; while if $c(xy) \in \{2, 3\}$, then define $\{c(xw), c(xz)\} = \{1, 4\}$. Then $c'(y) \neq c'(x)$. Furthermore, $c'(x) \neq c(xw)$ and $c'(x) \neq c(xz)$.

Case 1. $c'(y) = 1$. First, suppose that $\deg_T x = 2$ and x is adjacent to $w \in A_{k+1}$. If $c(xy) = 1$, then define $c(xw) = 2$; while if $c(xy) \in \{2, 3, 4\}$, then define $c(xw) = 1$. Thus $c'(x) \neq c'(y)$ and $c'(x) \neq c(xw)$. Next, suppose that $\deg_T x = 3$ and x is adjacent to the two vertices w and z in A_{k+1}. If $c(xy) = 1$, then define (i) $c(xw) = 2$ and $c(xz) = 4$ when w is not a leaf and (ii) $c(xw) = 0$ and $c(xz) = 2$ when w is a leaf. If $c(xy) \in \{2, 3\}$, then define $\{c(xw), c(xz)\} = \{1, 4\}$; while if $c(xy) = 4$, then define (i) $c(xw) = 1$ and $c(xz) = 2$ when z is not a leaf and (ii) $c(xw) = 3$ and $c(xz) = 0$ when z is a leaf. Then $c'(y) \neq c'(x)$. Furthermore, $c'(x) \neq c(xw)$ and $c'(x) \neq c(xz)$ when xw and xz are not pendant edges.

Case 2. $c'(y) = 2$. First, suppose that $\deg_T x = 2$ and x is adjacent to $w \in A_{k+1}$. If $c(xy) = 1$, then define $c(xw) = 2$; while if $c(xy) \in \{2, 3, 4\}$, then define $c(xw) = 1$. Thus $c'(x) \neq c'(y)$ and $c'(x) \neq c(xw)$. Next, suppose that $\deg_T x = 3$ and x is adjacent to the two vertices w and z in A_{k+1}. If $c(xy) \in \{1, 4\}$, then define $\{c(xw), c(xz)\} = \{2, 3\}$. If $c(xy) = 2$, then define (i) $c(xw) = 1$ and $c(xz) = 3$ when w is not a leaf and (ii) $c(xw) = 0$ and $c(xz) = 1$ when w is a leaf. If $c(xy) = 3$, then define $\{c(xw), c(xz)\} = \{1, 4\}$. Then $c'(y) \neq c'(x)$. Furthermore, $c'(x) \neq c(xw)$ and $c'(x) \neq c(xz)$ when xw and xz are not pendant edges.

Case 3. $c'(y) = 3$. First, suppose that $\deg_T x = 2$ and x is adjacent to $w \in A_{k+1}$. If $c(xy) \in \{1, 2\}$, then define $c(xw) = 3$; while if $c(xy) \in \{3, 4\}$, then define $c(xw) = 1$. Thus $c'(x) \neq c'(y)$ and $c'(x) \neq c(xw)$. Next, suppose that $\deg_T x = 3$ and x is adjacent to the two vertices w and z in A_{k+1}. If $c(xy) \in \{1, 4\}$, then define $\{c(xw), c(xz)\} = \{2, 3\}$. If $c(xy) = 2$, then define (i) $c(xw) = 1$ and $c(xz) = 3$ when w is not a leaf and (ii) $c(xw) = 4$ and $c(xz) = 1$ when w is a leaf. If $c(xy) = 3$, then define (i) $c(xw) = 1$ and $c(xz) = 2$ when w is not a leaf and

(ii) $c(xw) = 0$ and $c(xz) = 1$ when w is a leaf. Then $c'(y) \neq c'(x)$. Furthermore, $c'(x) \neq c(xw)$ and $c'(x) \neq c(xz)$ when xw and xz are not pendant edges.

Case 4. $c'(y) = 4$. First, suppose that $\deg_T x = 2$ and x is adjacent to $w \in A_{k+1}$. If $c(xy) \in \{1, 3\}$, then define $c(xw) = 2$; while if $c(xy) \in \{2, 4\}$, then define $c(xw) = 1$. Thus $c'(x) \neq c'(y)$ and $c'(x) \neq c(xw)$. Next, suppose that $\deg_T x = 3$ and x is adjacent to the two vertices w and z in A_{k+1}. If $c(xy) = 1$, then define $\{c(xw), c(xz)\} = \{2, 3\}$; while if $c(xy) \in \{2, 3\}$, then define $\{c(xw), c(xz)\} = \{1, 4\}$. If $c(xy) = 4$, then define (i) $c(xw) = 2$ and $c(xz) = 1$ when w is not a leaf and (ii) $c(xw) = 0$ and $c(xz) = 2$ when w is a leaf. Then $c'(y) \neq c'(x)$. Furthermore, $c'(x) \neq c(xw)$ and $c'(x) \neq c(xz)$ when xw and xz are not pendant edges.   □

Since $\chi_t'(K_{1,5}) = 7$ by Theorem 9.17, the upper bound of 7 for $\chi_t'(T)$ when T is a tree of maximum degree 5 cannot be improved.

Next, we verify Conjecture 9.16 for another class of trees. For an integer $r \geq 2$, a tree T is sometime called *r-regular* if each non-end-vertex of T has degree r. Thus, the degree set of an r-regular tree is $\{1, r\}$. In particular, a path P_n order $n \geq 3$ is 2-regular and a star $K_{1,r}$ is r-regular for $r \geq 2$. We first show that if T is an r-regular tree for some integer $r \geq 5$ such that $r \equiv 1 \pmod 4$, then $\chi_t'(T) \geq r + 2$. Some additional notation will be useful in the results that follow. For integers a and b with $a < b$, let

$$[a..b] = \{a, a+1, \ldots, b\}$$

be the set of integers between a and b, including a and b, and let $\sigma(a, b)$ denote the sum of the integers between a and b, that is,

$$\sigma(a, b) = \sum_{i=a}^{b} i = a + (a+1) + \cdots + b.$$

The following lemma is a useful observation.

Lemma 9.21. *Let $r \geq 5$ be an integer such that $r \equiv 1 \pmod 4$. Then*

$$\sigma(0, r) - j \not\equiv j \pmod{r+1}$$

for each integer $j \in [0..r]$.

Theorem 9.22. *If T is a regular tree of order at least* 6 *such that $\Delta(T) \equiv 1$ (mod 4), then*

$$\chi_t'(T) \geq \Delta(T) + 2.$$

Proof. By Theorem 9.17, we may assume that T is not a star. Suppose that T is an r-regular tree for some integer $r \geq 5$ and $r \equiv 1 \pmod 4$. Then $\Delta(T) = r$. By Observation 9.15, it follows that $\chi_t'(T) \geq r + 1$. We first show that $\chi_t'(T) \neq r + 1$. Assume, to the contrary, that $\chi_t'(T) = r + 1$. Let $c : E(T) \to \mathbb{Z}_{r+1}$ be a twin edge $(r + 1)$-coloring of T. Let $v_1 \in V(T)$ such that $\deg v_1 = r$. Then there is exactly one color in $\mathbb{Z}_{r+1}$ that is not assigned to any edge incident with v_1 by c. Suppose that $\{c(v_1 w) : w \in N(v_1)\} = \mathbb{Z}_{r+1} - \{j_1\}$ for some integer j_1; that is, j_1 is the only color that is not assigned to any edge incident with v_1. Since $c'(v_1) = \sigma(0, r) - j_1$ and $r \equiv 1 \pmod 4$, it follows by Lemma 9.21 that $c'(v_1) \neq j_1$ and so $c'(v_1) \in \mathbb{Z}_{r+1} - \{j_1\}$. Hence $c'(v_1) = c(v_1 w)$ for some $w \in N(v_1)$. Let $c'(v_1) = c(v_1 v_2)$, where $v_2 \in N(v_1)$. If v_2 is an end-vertex of T, then $c'(v_2) = c(v_1 v_2) = c'(v_1)$, which is impossible. Thus v_2 is not an end-vertex of T and so $\deg v_2 = r$. Suppose that $\{c(v_2 w) : w \in N(v_2)\} = \mathbb{Z}_{r+1} - \{j_2\}$ for some integer j_2, where then $c(v_1 v_2) \in \mathbb{Z}_{r+1} - \{j_2\}$. By Lemma 9.21 again, $c'(v_2) \neq j_2$ and so $c'(v_2) = c(v_2 v_3)$ for some $v_3 \in N(v_2)$. Since c is a twin edge $(r + 1)$-coloring of T, it follows that $c'(v_2) \neq c'(v_1) = c(v_1 v_2)$; which implies that $v_3 \neq v_1$. A similar argument shows that v_3 is not an end-vertex and so $\deg v_3 = r$. Continuing in this manner, we arrive at a sequence $v_1, v_2, \ldots, v_k$ of $k \geq 2$ distinct vertices of degree r in T such that (1) $v_i v_{i+1} \in E(T)$ for $1 \leq i \leq k - 1$ and (2) v_{k-1} is the only non-end-vertex to which v_k is adjacent and so v_k is adjacent to exactly $r - 1$ end-vertices of T. Suppose that $\{c(v_k w) : w \in N(v_k)\} = \mathbb{Z}_{r+1} - \{j_k\}$ for some integer j_k, where then $c(v_{k-1} v_k) \in \mathbb{Z}_{r+1} - \{j_k\}$. It then follows by Lemma 9.21 that $c'(v_k) \neq j_k$. Hence $c'(v_k) = c(v_k v_{k+1})$ for some $v_{k+1} \in N(v_k) - \{v_{k-1}\}$. Since v_{k+1} is an end-vertex of T, it follows that $c'(v_{k+1}) = c(v_k v_{k+1}) = c'(v_k)$, which is impossible. Therefore, $\chi'_t(T) \neq r + 1$ and so $\chi'_t(T) \geq r + 2$. $\qquad\square$

First, we determine the twin chromatic indices of all regular double stars. Recall that if the central vertices of a double star have degrees a and b, respectively, then it is denoted by $S_{a,b}$. Thus, if $a = b$, then $S_{a,b}$ is a *regular* double star.

Theorem 9.23 ([8]). *If T is a regular double star, then*

$$
\chi'_t(T) = \begin{cases} \Delta(T) + 1 & \text{if } \Delta(T) \not\equiv 1 \pmod 4 \\ \Delta(T) + 2 & \text{if } \Delta(T) \equiv 1 \pmod 4 \end{cases}
$$

Proof. By Theorem 9.17, we may assume that T is not a path. Let $T = S_{r,r}$ for some integer r, where then $\Delta(T) = r \geq 3$. Suppose that the central vertices are u and v, where $\deg u = \deg v = r$. Let $u_1, u_2, \ldots, u_{r-1}$ be the end-vertices of T that are adjacent to u and let $v_1, v_2, \ldots, v_{r-1}$ be the end-vertices of T that are adjacent to v. First, suppose that $r \not\equiv 1 \pmod 4$. By Observation 9.15, it follows that $\chi'_t(T) \geq r + 1$. It remains to show that T has a twin $(r + 1)$-edge coloring $c : E(T) \to \mathbb{Z}_{r+1}$. We consider two cases, according to whether r is even or r is odd.

Case 1. $r \geq 4$ *is even.* Then $\sigma(1, r) = 0$ in $\mathbb{Z}_{r+1}$. Define $c(uv) = r$ and

$$
\{c(uu_i) : 1 \leq i \leq r - 1\} = [1..r - 1]
$$

$$
\{c(vv_i) : 1 \leq i \leq r - 1\} = [0..r - 1] - \{1\}.
$$

Then c is a proper edge coloring. Observe that $c'(u) = \sigma(1, r) = 0$ in $\mathbb{Z}_{r+1}$ and $c'(v) = \sigma(1, r) - 1 = r$ in $\mathbb{Z}_{r+1}$. Thus $c'(u) \neq c'(v)$. Furthermore, $c'(u_i) \neq 0 = c'(u)$ and $c'(v_i) \neq r = c'(v)$ for $1 \leq i \leq r - 1$. Hence, c' is a proper vertex coloring and so c is a twin $(r + 1)$-edge coloring of T.

Case 2. $r \geq 3$ *is odd.* Since $r \not\equiv 1 \pmod 4$, it follows that $r \equiv 3 \pmod 4$ and so $r = 4t+3$ for some integer $t \geq 0$. First, suppose that $t = 0$. Define $c : E(T) \to \mathbb{Z}_4$ by $c(uv) = 2$, $\{c(uu_1), c(uu_2)\} = \{0, 1\}$ and $\{c(vv_1), c(vv_2)\} = \{1, 3\}$. Then $c'(u) = 3 \neq c'(u_i)$ in $\mathbb{Z}_4$ and $c'(v) = 2 \neq c'(v_i)$ in $\mathbb{Z}_4$ for $i = 1, 2$. Also, $c'(u) \neq c'(v)$. Therefore, c is a twin edge 4-coloring of T.

Next, suppose that $t \geq 1$. Observe that $\sigma(0, 4t + 3) = 2t + 2$ in $\mathbb{Z}_{r+1} = \mathbb{Z}_{4t+4}$ and so

$$\sigma(0, 4t + 3) - (t + 1) = t + 1 \text{ in } \mathbb{Z}_{4t+4}. \tag{9.6}$$

Define $c(uv) = 2t + 2$ and

$$\{c(uu_i) : 1 \leq i \leq 4t + 2\} = [0..4t + 3] - \{t + 1, 2t + 2\}$$

$$\{c(vv_i) : 1 \leq i \leq 4t + 2\} = [1..4t + 3] - \{2t + 2\}.$$

Then c is a proper edge coloring of T. By (9.6),

$$c'(u) = \sigma(1, 4t + 3) - (t + 1) = t + 1 \text{ in } \mathbb{Z}_{4t+4}$$

$$c'(v) = \sigma(1, 4t + 3) = 2t + 2 \text{ in } \mathbb{Z}_{4t+4}.$$

Thus, $c'(u) \neq c'(v)$. Furthermore, $c'(u_i) \neq t + 1 = c'(u)$ in $\mathbb{Z}_{4t+4}$ and $c'(v_i) \neq 2t + 2 = c'(v)$ in $\mathbb{Z}_{4t+4}$ for $1 \leq i \leq 4t + 2$. Hence, c' is a proper vertex coloring and so c is a twin $(r + 1)$-edge coloring of T.

Next, suppose that $r \geq 5$ and $r \equiv 1 \pmod 4$. By Theorem 9.22, it follows that $\chi'_t(T) \geq r + 2$. Thus, it remains to show that T has a twin edge $(r + 2)$-coloring $c : E(T) \to \mathbb{Z}_{r+2}$. Let $r = 4t + 1$ for some integer $t \geq 1$. Then $\sigma(0, 4t + 2) = 0$ in $\mathbb{Z}_{r+2} = \mathbb{Z}_{4t+3}$ and so

$$\sigma(0, 4t + 2) - (2t + 1) - 1 = 2t + 1 \text{ in } \mathbb{Z}_{4t+3}$$

$$\sigma(0, 4t + 2) - (2t) - 3 = 2t \text{ in } \mathbb{Z}_{4t+3}$$

Define $c(uv) = 2t + 2$ and

$$\{c(uu_i) : 1 \leq i \leq 4t\} = [0..4t + 2] - \{1, 2t + 1, 2t + 2\}$$

$$\{c(vv_i) : 1 \leq i \leq 4t\} = [0..4t + 2] - \{3, 2t, 2t + 2\}.$$

Fig. 9.5 A twin edge
7-coloring of $S_{5,5}$

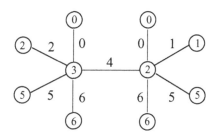

Then c is a proper edge coloring of T. Observe that

$$c'(u) = \sigma(0, 4t + 2) - (2t + 1) - 1 = 2t + 1 \text{ in } \mathbb{Z}_{4t+3}$$

$$c'(v) = \sigma(0, 4t + 2) - 2t - 3 = 2t \text{ in } \mathbb{Z}_{4t+3},$$

Thus, $c'(u) \neq c'(v)$. Furthermore, $c'(u_i) \neq 2t + 1 = c'(u)$ and $c'(v_i) \neq 2t = c'(v)$ for $1 \leq i \leq 4t$. The colorings c and c' are shown for $r = 5$ in Fig. 9.5. Hence, c' is a proper vertex coloring and so c is a twin edge $(r + 2)$-coloring of T. □

It was shown in [8] that for every double star T whose central vertices have different degrees, there is a twin edge $(\Delta(T) + 1)$-coloring of T.

Theorem 9.24 ([8]). *If T is a double star whose central vertices have different degrees, then*

$$\chi'_t(T) \leq \Delta(T) + 1.$$

The following is a consequence of Theorems 9.22, 9.23 and 9.24.

Corollary 9.25. *A double star T has $\chi'_t(T) = \Delta(T) + 2$ if and only if T is an r-regular tree for some integer $r \geq 5$ with $r \equiv 1 \pmod 4$.*

In general, Conjecture 9.16 has been verified for regular trees.

Theorem 9.26 ([8]). *If T is a regular tree of order at least 3, then*

$$\chi'_t(T) \leq \Delta(T) + 2.$$

By Theorems 9.22 and 9.26, if T is a regular tree of order at least 6 such that $\Delta(T) \equiv 1 \pmod 4$, then $\chi'_t(T) = \Delta(T) + 2$. Furthermore, we saw in Corollary 9.25 that if T is a double star, then $\chi'_t(T) = \Delta(T) + 2$ if and only if T is an r-regular tree for some integer $r \geq 5$ with $r \equiv 1 \pmod 4$ where then $r = \Delta(T)$. From the examples we are aware of, it suggests that Corollary 9.25 is true for all trees in general. In any case, this problem appears to be worthy of further study.

References

1. Addario-Berry, L., Aldred, R. E. L., Dalal, K., & Reed, B. A. (2005). Vertex colouring edge partitions. *The Journal of Combinatorial Theory, 94,* 237–244.
2. Aigner, M., & Triesch, E. (1990). Irregular assignments and two problems á la Ringel. In R. Bodendiek & R. Henn (Eds.), *Topics in combinatorics and graph theory* (pp. 29–36). Heidelberg: Physica.
3. Aigner, M., & Triesch, E. (1990). Irregular assignments of trees and forests. *The SIAM Journal on Discrete Mathematics, 3,* 439–449.
4. Aigner, M., Triesch, E., & Tuza, Z. (1992). Irregular assignments and vertex-distinguishing edge-colorings of graphs. In *Combinatorics' 90* (pp. 1–9). New York: Elsevier Science.
5. Albertson, M., & Collins, K. (1996). Symmetric breaking in graphs. *The Electronic Journal of Combinatorics, 3,* R18.
6. Amar, D., & Togni, O. (1998). Irregularity strength of trees. *Discrete Mathematics, 190,* 15–38.
7. Andrews, E., Johnston, D., & Zhang, P. (2014). A twin edge coloring conjecture. *Bulletin of the Institute of Combinatorics and Its Applications, 70,* 28–44.
8. Andrews, E., Johnston, D., & Zhang, P., On twin edge colorings in trees. *Journal of Combinatorial Mathematics and Combinatorial Computing* (to appear).
9. Andrews, E., Helenius, L., Johnston, D., VerWys, J., & Zhang, P. (2014). On twin edge colorings of graphs. *The Discussiones Mathematicae Graph Theory, 34,* 613–627.
10. Balister, P. N., Györi, E., Lehel, J., & Schelp, R. H. (2007). Adjacent vertex distinguishing edge-colorings. *The SIAM Journal on Discrete Mathematics, 21,* 237–250.
11. Baril, J. L., Kheddouci, H., & Togni, O. (2005). The irregularity strength of circulant graphs. *Discrete Mathematics, 304,* 1–10.
12. Bazgan, C., Harkat-Benhamdine, A., Li, H., & Woźniak, M. (1999). On the vertex-distinguishing proper edge-colorings of graphs. *Journal of Combinatorial Theory, Series B, 75,* 288–301.
13. Bohman, T., & Kravitz, D. (2004). On the irregularity strength of trees. *Journal for Graph Theory, 45,* 241–254.
14. Bondy, J. A., & Chvátal, V. (1976). A method in graph theory. *Discrete Mathematics, 15,* 111–136.
15. Brooks, R. L. (1941). On coloring the nodes of a network. *Proceedings of the Cambridge Philosophical Society, 37,* 194–197.
16. Burris, A. C. (1994). On graphs with irregular coloring number 2. *Congressus Numerantium, 100,* 129–140.

© Ping Zhang 2015
P. Zhang, *Color-Induced Graph Colorings*, SpringerBriefs in Mathematics,
DOI 10.1007/978-3-319-20394-2

17. Burris, A. C. (1995). The irregular coloring number of a tree. *Discrete Mathematics, 141,* 279–283.
18. Burris, A. C., & Schelp, R. H. (1997). Vertex-distinguishing proper edge colorings. *Journal for Graph Theory, 26,* 73–82.
19. Chartrand, G., & Zhang, P. (2009). *Chromatic graph theory.* Boca Raton: Chapman & Hall/CRC.
20. Chartrand, G., & Zhang, P. (2011). *Discrete mathematics,* Waveland Press, Long Grove, IL.
21. Chartrand, G., English, S., & Zhang, P., Binomial colorings of graphs. Preprint.
22. Chartrand, G., Lesniak, L., & Zhang, P. (2010). *Graphs & digraphs* (5th ed.). Boca Raton, FL: Chapman & Hall/CRC.
23. Chartrand, G., Escuadro, H., Okamoto, F., & Zhang, P. (2006). Detectable colorings of graphs. *Utilitas Mathematica, 69,* 13–32.
24. Chartrand, G., Jacobson, M. S., Lehel, J., Oellermann, O. R., Ruiz, S., & Saba, F. (1988). Irregular networks. *Congressus Numerantium, 64,* 197–210.
25. Cuckler, B., & Lazebnik, F. (2008). Irregularity strength of dense graphs. *Journal for Graph Theory, 58,* 299–313.
26. Dinitz, J. H., Garnick, D. K., & Gyárfás, A. (1992). On the irregularity strength of the $m \times n$ grid. *Journal for Graph Theory, 16,* 355–374.
27. Dirac, G. A. (1952). Some theorems on abstract graphs. *Proceedings of the London Mathematical Society, 2,* 69–81.
28. Dong, A. J., Wang, G. H., & Zhang, J. H. (2014). Neighbor sum distinguishing edge colorings of graphs with bounded maximum average degree. *Discrete Applied Mathematics, 166,* 84–90.
29. Ebert, G., Hemmeter, J., Lazebnik, F., & Woldar, A. (1990). Irregularity strengths for certain graphs. *Congressus Numerantium, 71,* 39–52.
30. Ebert, G., Hemmeter, J., Lazebnik, F., & Woldar, A. (1991). On the number of irregular assignments on a graph. *Discrete Mathematics, 93,* 131–142.
31. Entringer, R. C., & Gassman, L. D. (1974). Line-critical point determining and point distinguishing graphs. *Discrete Mathematics, 10,* 43–55.
32. Escuadro, H. (2006). *Detectable colorings of graphs.* Ph.D. Dissertation, Western Michigan University.
33. Escuadro, H., & Zhang, P. (2005). On detectable colorings of graphs. *Mathematica Bohemica, 130,* 427–445.
34. Escuadro, H., & Zhang, P. (2005). Extremal problems on detectable colorings of connected graphs with cycle rank 2. *AKCE International Journal of Graphs and Combinatorics, 2,* 99–117.
35. Escuadro, H., & Zhang, P. (2008). Extremal problems on detectable colorings of trees. *Discrete Mathematics, 308,* 1951–1961.
36. Escuadro, H., Okamoto, F., & Zhang, P. (2006). On detectable factorizations of cubic graphs *Journal of Combinatorial Mathematics and Combinatorial Computing, 56,* 47–63.
37. Escuadro, H., Okamoto, F., & Zhang, P. (2008). A three-color problem in graph theory. *Bulletin of the Institute of Combinatorics and Its Applications, 52,* 65–82.
38. Faudree, R. J., & Lehel, J. (1987). Bound on the irregularity strength of regular graphs. In *Combinatorics. Colloq. Math. Soc. János Bolyai* (Vol. 52, pp. 247–256). Amsterdam: North Holland.
39. Faudree, R. J., Gyárfás, A., & Schelp, R. H. (1987). On graphs of irregularity strength 2. In *Combinatorics. Colloq. Math. Soc. János Bolyai* (Vol. 52, pp. 239–246). Amsterdam: North Holland.
40. Faudree, R. J., Jacobson, M. S., Kinch, L., & Lehel, J. (1991). Irregularity strength of dense graphs. *Discrete Mathematics, 91,* 45–59.
41. Faudree, R. J., Jacobson, M. S., Lehel, J., & Schelp, R. H. (1989). Irregular networks, regular graphs and integer matrices with distinct row and column sums. *Discrete Mathematics, 76,* 223–240.
42. Flandrin, E., Marczyk, A., Przybylo, J., Saclé, J. F., & Woźniak, M. (2013). Neighbor sum distinguishing index. *Graphs and Combinatorics, 29,* 1329–1336.

43. Fournier, J.-C. (1973). Colorations des arétes d'un graphe. In *Colloque sur la Théorie des Graphes (Brussels, 1973)* (French). Cahiers Centre Études Recherche Opér (Vol. 15, pp. 311–314).

44. Frank, O., Harary, F., & Plantholt, M. (1982). The line-distinguishing chromatic number of a graph. *Ars Combinatoria, 14*, 241–252.

45. Frieze, A., Gould, R., Karoński, M., & Pfender, F. (2002). On graph irregularity strength. *Journal for Graph Theory, 41*, 120–137.

46. Fujie-Okamoto, F., & Will, T. G. (2012). Efficient computation of the modular chromatic numbers of trees. *Journal of Combinatorial Mathematics and Combinatorial Computing, 82*, 77–86.

47. Gallian, J. A. (2009). A dynamic survey of graph labeling. *The Electronic Journal of Combinatorics, 16*, #DS6.

48. Gnana Jothi, R. B. (1991). *Topics in graph theory*. Ph.D. Thesis, Madurai Kamaraj University.

49. Golomb, S. W. (1972). How to number a graph. In *Graph theory and computing* (pp. 23–37). New York: Academic.

50. Graham, R. L., & Sloane, N. J. A. (1980). On addition bases and harmonious graphs. *The SIAM Journal on Discrete Mathematics, 1*, 382–404.

51. Gyárfás, A. (1988). The irregularity strength of $K_{m,m}$ is 4 for odd m. *Discrete Mathematics, 71*, 273–274.

52. Gyárfás, A. (1989). The irregularity strength of $K_n - mK_2$. *Utilitas Mathematica, 35*, 111–113.

53. Gyárfás, A., Jacobson, M. S., Kinch, L., Lehel, J., & Schelp, R. H. (1992). Irregularity strength of uniform hypergraphs. *Journal of Combinatorial Mathematics and Combinatorial Computing, 11*, 161–172.

54. Győri, E., Horňák, M., Palmer, C., & Woźniak, M. (2008). General neighbor-distinguishing index of a graph. *Discrete Mathematics, 308*, 827–831.

55. Harary, F., & Plantholt, M. (1983). Graphs with the line-distinguishing chromatic number equal to the usual one. *Utilitas Mathematica, 23*, 201–207.

56. Harary, F., & Plantholt, M. (1985). The point-distinguishing chromatic index. In *Graphs and applications* (pp. 147–162). New York: Wiley.

57. Hopcroft, J. E., & Krishnamoorthy, M. S. (1983). On the harmonious coloring of graphs. *SIAM Journal on Algebraic Discrete Methods, 4*, 306–311.

58. Johnston, D., & Zhang, P. (2014). An upper bound for the twin chromatic index of a graph. *Congressus Numerantium, 219*, 175–182.

59. Jones, R. (2011). *Modular and graceful edge colorings of graphs*. Ph.D. Dissertation, Western Michigan University.

60. Jones, R., & Zhang, P. (2012). Nowhere-zero modular edge-graceful graphs. *The Discussiones Mathematicae Graph Theory, 32*, 487–505.

61. Jones, R., Kolasinski, K., & Zhang, P. (2012). A proof of the modular edge-graceful trees conjecture. *Journal of Combinatorial Mathematics and Combinatorial Computing, 80*, 445–455.

62. Jones, R., Kolasinski, K., Okamoto, F., & Zhang, P. (2011). Modular neighbor-distinguishing edge colorings of graphs. *Journal of Combinatorial Mathematics and Combinatorial Computing, 76*, 159–175.

63. Jones, R., Kolasinski, K., Okamoto, F., & Zhang, P. (2012). On modular chromatic indexes of graphs. *Journal of Combinatorial Mathematics and Combinatorial Computing, 82*, 295–306.

64. Jones, R., Kolasinski, K., Okamoto, F., & Zhang, P. (2013). On modular edge-graceful graphs. *Graphs and Combinatorics, 29*, 901–912.

65. Kalkowski, M., Karoński, M., & Pfender, F. (2010). Vertex-coloring edge-weightings: towards the 1-2-3 Conjecture. *Journal of Combinatorial Theory, Series B, 100*, 347–349.

66. Kalkowski, M., Karoński, M., & Pfender, F. (2011). A new upper bound for the irregularity strength of graphs. *The SIAM Journal on Discrete Mathematics, 25*, 1319–1321.

67. Karoński, M., Łuczak, T., & Thomason, A. (2004). Edge weights and vertex colours. *The Journal of Combinatorial Theory, 91*, 151–157.

68. Kinch, L., & Lehel, J. (1991). The irregularity strength of tP_3. *Discrete Mathematics, 94,* 75–79.

69. König, D. (1916). Über Graphen und ihre Anwendung auf Determinantentheorie und Mengenlehre. *Mathematische Annalen, 77,* 453–465

70. Mahéo, M., & Saclé, J. F. (2008). Some results on $(\sum, p, g)$-valuation of connected graphs. Report de Recherche 1497. Université de Paris-Sud, Center d'Orsay.

71. Nierhoff, T. (2000). A tight bound on the irregularity strength of graphs. *The SIAM Journal on Discrete Mathematics, 13,* 313–323.

72. Petersen, J. (1891). Die Theorie der regulären Graphen. *Acta Mathematica, 15,* 193–220.

73. Przybylo, J. (2008). Irregularity strength of regular graphs. *The Electronic Journal of Combinatorics, 15,* #R82.

74. Przybylo, J., & Woźniak, M. (2010). On a 1, 2 Conjecture. *Discrete Mathematics & Theoretical Computer Science, 12,* 101–108.

75. Rosa, A. (1967). On certain valuations of the vertices of a graph. In *Theory of Graphs, Proceedings of International Symposium, Rome 1966* (pp. 349–355). New York: Gordon and Breach.

76. Tait, P. G. (1880). Remarks on the colouring of maps. *Proceedings of the Royal Society of Edinburgh, 10,* 501–503, 729.

77. Togni, O. (2000). Irregularity strength and compound graphs. *Discrete Mathematics, 218,* 235–243.

78. Vizing, V. G. (1964). On an estimate of the chromatic class of a p-graph. *Diskret Analiz, 3* (Russian), 25–30.

79. Wang, G. H., & Yan, G. Y. (2014). An improved upper bound for the neighbor sum distinguishing index of graphs. *Discrete Applied Mathematics, 175,* 126–128.

80. Wang, G. H., Chen, Z. M., & Wang, J. H. (2014). Neighbor sum distinguishing index of planar graphs. *Discrete Mathematics, 334,* 70–73.

81. Zhang, Z., Liu, L., & Wang, J. (2002). Adjacent strong edge coloring of graphs. *Applied Mathematics Letters, 15,* 623–626.

Index

© Ping Zhang 2015
P. Zhang, *Color-Induced Graph Colorings*, SpringerBriefs in Mathematics,
DOI 10.1007/978-3-319-20394-2

Printed in the United States
By Bookmasters